T0140716

"Tectonic evolution of the Lake Ohrid Basin (Macedonia/Albania)"

Von der Fakultät für Georessourcen und Materialtechnik

der Rheinisch-Westfälischen Technischen Hochschule Aachen

zur Erlangung des akademischen Grades eines

Doktors der Naturwissenschaften

genehmigte Dissertation

vorgelegt von

M.A. Nadine Hoffmann

aus Dachau

Berichter: **Univ.-Prof. Dr. rer. nat. Klaus Reicherter**
Ass. Prof. Dr. Alessandro M. Michetti

Tag der mündlichen Prüfung 18. Juli 2013

Bibliografische Information der Deutschen Nationalbibliothek

Die Deutsche Nationalbibliothek verzeichnet diese Publikation in der Deutschen Nationalbibliografie; detaillierte bibliografische Daten sind im Internet über http://dnb.d-nb.de abrufbar.

ISBN 978-3-8325-3586-5

Logos Verlag Berlin GmbH
Comeniushof, Gubener Str. 47,
10243 Berlin
Tel.: +49 (0)30 42 85 10 90
Fax: +49 (0)30 42 85 10 92
INTERNET: http://www.logos-verlag.de

“Man darf nie an die ganze Straße auf einmal denken [...] man muss nur an den nächsten Schritt denken, an den nächsten Atemzug, an den nächsten Besenstrich. Und immer wieder nur an den nächsten. Auf einmal merkt man, dass man Schritt für Schritt die ganze Straße gemacht hat. Man hat gar nicht gemerkt wie, und man ist nicht außer Puste.”

Beppo Straßenkehrer

Abstract

Lake Ohrid is located at the border between the Republic of Macedonia and the Republik of Albania and is situated in an active tectonic region of the Balkanides. Several phases of deformation affected the area since the Tertiary which is today mainly controlled by the influence of the Northern Hellenic Trench and the North Anatolian Fault Zone. This results in zonation into a compressional coastal domain, separated by a narrow zone of transition to the extensional domain further east in which the Neogene basins formed. The Lake Ohrid Basin forms one of the most active seismic zones in Albania/Macedonia, which is documented by several moderate earthquakes in the last few centuries and a few major earthquakes during historical times. This seismic activity has created a seismic landscape with a variety of morphological features that are preserved in the surroundings. Earthquake focal mechanisms show active N-S normal faulting with horst and graben structures in a basin and range like environment. This study provides insight into landscape formation and the landforming processes that the Ohrid Basin has been exposed to since the Quaternary. A multidisciplinary approach with a combination of different geoscientific methods (e.g. structural and palaeostress analysis, evaluation of the historical and instrumental seismicity, shallow geophysics, remote sensing) was chosen. According to this a workflow is proposed for neotectonic studies in the Mediterranean and areas with similar climatic conditions.

The main objectives were: to determine the palaeostress fields which controlled the evolution of the area; to gain information on fault orientation, fault geometry and spatial distribution of fault scarps; to determine the grade of activity in the Lake Ohrid Basin; and to define active/less active areas experiencing tectonic deformation.

The palaeostress analysis revealed major shifts in the geodynamic setting, which can be described by three successive periods of basin development: (1) an orogenic phase with (i) NE-SW compression in Cretaceous-Paleogene, (ii) NE-SW extension in Late Eocene-Oligocene, and (iii) NE-SW shortening and strike slip movement in Oligocene-Miocene; (2) a transtensional phase with NW-SE extension and strike-slip movement in Mid-Miocene; and (3) E-W extension from Late Miocene to present.

Sedimentological studies with data from shallow drillings and geophysical investigations provided insight into the Holocene coastal evolution of the lake and allowed two main geomorphological systems to be determined. The plains north and south of the lake are dominated by clastic input related to climate variations and uplift/erosion. In contrast, the steep western and eastern margins are controlled by recent tectonics and normal faulting.

Geomorphological studies show that the lithological zonation of the basin causes an inhomogeneous morphological surface expression within the influence of the same stress field. The main trend of the normal faults is N-S and therefore a graben structure has formed. The average geometry of the fault scarps measured in the field is an upper slope that dips at 22°, beneath this the scarp's free face dip angle ranges between 42° and 85° and a lower slope that matches the upper slope angle at 22° in most cases. The mean constructed fault height is 39.5 m for the west coast, 29.3 m for the east coast, and 17.6 m for the north. Fault lengths vary between 10 and 20 km and are expected to have the potential for earthquakes between M 6.5 - 7.0. The values of calculated slip rates inferring a post-glacial development of the exposed fault scarps range between 0.28 and 3.25 mm/a for the last 18 ka. These values far exceed well defined slip rates in comparable conditions. This leads to the concept of "stop and go" faults where the older outer faults slip every time a younger fault evolves. This therefore creates a higher relief.

The fault scarps are also influenced by gravitational forces which cause the highly fractured rocks to react to seismic events, or large offshore mass wasting processes, and slip downslope. Mainly, the west coast is dominated by mass wasting processes, while the east coast is highly segmented with tilted blocks of basement.

This study has illustrated that by combining different fieldwork techniques our knowledge of the seismicity and tectonic evolution of an area can be significantly advanced.

Zusammenfassung

Der Ohrid See liegt an der Grenze zwischen der Republik Mazedonien und der Republik Albanien und damit in einer tektonisch aktiven Region der Balkaniden. Mehrere Deformationsphasen beeinflussten das Gebiet seit dem Tertiär. Heute steht es hauptsächlich unter dem Einfluss des Nordhellenischen Grabens und der Nordanatolischen Störung. Dadurch ergibt sich eine Zonierung, der Region in ein kompressiv dominiertes Küstengebiet, das nach Osten hin durch eine schmale Übergangszone von der extensiven Zone getrennt ist, in der sich die Neogenen Becken entwickelt haben. Das Ohrid Becken ist eines der seismisch aktivsten Gebiete im albanisch/mazedonischen Raum. Dies zeigt sich in einigen moderaten Erdbeben der letzten Jahrhunderte und in wenigen großen Erdbeben die in historischen Zeiten dokumentiert wurden. Diese seismische Aktivität führte zur Entwicklung einer seismischen Landschaft mit einem großen Formenschatz. Herdflächenlösungen rezenter Erdbeben zeigen aktive N-S verlaufende Abschiebungen, die sich in Horst und Graben Strukturen manifestieren und Ausdruck einer Basin und Range Provinz sind. Die vorliegende Arbeit gibt einen Einblick in die Entwicklung der Landschaft und in die Prozesse die das Gebiet seit dem Quartär beeinflussen. Ein multidisziplinärer Ansatz mit einer Kombination verschiedenener geowissenschaftlicher Methoden (z.B. Struktur- und Paläostressanlyse, Auswertung der historischen und instrumentellen Erdbebendaten, geophysikalische Methoden, Fernerkundung) wurde gewählt. Daraus wurde ein exemplarischer Workflow für neotektonische Studien im Mittelmeerraum beziehungsweise für vergleichbare klimatische Bedingungen entwickelt.

Die wichtigsten Ziele, die zu Beginn der Studie definiert wurden, sind: die Paläostressfelder zu rekonstruieren, die die Entwicklung der Region beeinflusst haben, Informationen über Störungsorientierung, -geometrie und räumliche Verbreitung der Störungsstufen zu erhalten, den Aktivitätsgrad des Ohrid Beckens zu bestimmen und Gebiete höherer bzw. geringerer Aktivität auszuweisen, die unter dem Einfluss tektonischer Deformation stehen.

Die Paläostressanalyse deckte große Veränderung innerhalb des geodynamischen Systems auf, die durch fünf aufeinanderfolgende Perioden der Beckenentwicklulng beschrieben werden können: (1) Orogene Phase mit (i) NE-SW Kompression in der Kreide-Paläogen; (ii) NE-SW Extension im späten Eozän-Oligozän; (iii) NE-SW Kompression und Blattverschiebungen im Oligozän-Miozän; (2) Transtensionsphase mit NW-SE Extension und Blattverschiebungen im mittleren Miozän; (3) E-W Extension seit dem späten Miozän bis heute.

Sedimentologische Untersuchungen anhand von Daten aus Flachbohrungen und geophysikalischen Messungen geben einen Einblick in die Holozäne Küstenentwicklung. Daraus resultiert die Differenzierung in zwei geomorphologische Systeme: Die Ebenen im Norden und Süden des Sees, die durch klastischen Eintrag im Zusammenhang mit klimatischen

Veränderungen und Uplift/Erosion dominiert sind und im Kontrast dazu die steilen West- und Ostflanken, die im Wesentlichen durch die aktuelle Tektonik und den daraus resultierenden Abschiebungen beeinflusst sind.

Geomorphologische Studien zeigen dass die lithologische Zonierung des Beckens eine differenzierte morphologische Landschaftsentwicklung unter dem Einfluss desselben Stressfeldes zur Folge hat. Abschiebungen streichen hauptsächlich N-S und bilden einen Graben aus. Die Geometrie der Störungsstufen, die im Geände gemessen wurden, wird definiert durch einen oberen Hang mit einer durchschnittlichen Neigung von 22°, der darunterliegenden Störungsfläche, die mit einem Winkel zwischen 42° und 85° einfällt und dem unteren Hang mit einem Einfallwinkel von 22°, was in den meisten Fällen dem Winkel des oberen Hangs entspricht. Die durchschnittliche Höhe der Störungsstufen beträgt 39.5 m an der Westküste; 29.3 m an der Ostküste; und 17.6 m im Norden. Die Länge der Störungen liegt in der Regel zwischen 10 und 20 km. Diese sind in der Lage Erdbeben mit einer Magnitude M 6.5 - 7.0 zu generieren. Die Werte der berechneten Slipraten unter der Annahme einer postglazialen Entwicklung der Störungen liegen zwischen 0.28 und 3.25 mm/a für die letzten 18 ka. Diese Werte überschreiten bei Weitem gut dokumentierte Slipraten unter vergleichbaren Bedingungen. Was zu dem Konzept der "stop and go" Störungen führt, bei dem die äußere und ältere Störung sich jedes Mal mitbewegt, sobald sich eine jüngere Störung entwickelt; dadurch kann sich ein markanteres Relief ausbilden.

Die Störungsstufen werden darüber hinaus durch gravitationsbedingte Massenbewegungen beeinflusst, die ein Abrutschen der deformierten Gesteine in Verbindung mit Erdbebenereignissen zur Folge haben.

Die vorliegende Arbeit konnte anschaulich machen, dass durch die Kombination verschiedener Geländemethoden unser Wissen um die Seismizität und die tektonische Entwicklung eines Gebietes signifikant verbessert werden kann.

Contents

Abstract V

Zusammenfassung VII

1 Introduction 1

1.1 Scope of the thesis . . . 3
1.2 Structure of the thesis . . . 5

2 Methods 7

2.1 Palaeostress Analysis . . . 7
2.2 Geophysical Techniques . . . 11
2.2.1 Ground Penetrating Radar . . . 11
2.2.2 Electrical Resistivity . . . 12
2.2.3 Sidescan Sonar . . . 13
2.3 Percussion drilling . . . 13
2.3.1 X-ray fluorescence analysis . . . 14
2.3.2 Magnetic Susceptibility . . . 15
2.3.3 Combining geophysical methods across faults . . . 15
2.4 Dating methods . . . 16
2.4.1 Radiocarbon dating . . . 16
2.4.2 OSL dating . . . 17
2.5 Tectonic morphology studies . . . 17
2.5.1 Remote Sensing . . . 17
2.5.2 Geomorphological Analysis . . . 19
2.5.3 Workflow . . . 21

3 Regional Setting 23

3.1 Geodynamic setting . . . 23
3.2 Seismicity and Neotectonics . . . 28
3.3 Regional Geology . . . 32
3.3.1 Korabi Zone . . . 34
3.3.2 Mirdita Zone . . . 36
3.3.3 Syn- and postorogenic development . . . 36
3.3.4 Hydrology . . . 38

4 Palaeostress Analysis 41

4.1 Orogenic Phase . . . 44
4.2 Transtensional Phase . . . 51
4.3 Extensional Phase . . . 53
4.4 Discussion . . . 54

5 Sedimentological Investigations 59
5.1 Struga . . . 62
5.1.1 Drilling and core logging . . . 62
5.1.2 Geophysical investigations . . . 64
5.1.3 Interpretation . . . 64
5.2 Velestovo . . . 66
5.2.1 Drilling and core logging . . . 66
5.2.2 Geophysical investigations . . . 69
5.2.3 Interpretation . . . 70
5.3 Sveti Naum . . . 72
5.3.1 Drilling and core logging . . . 72
5.3.2 Geophysical investigations . . . 74
5.3.3 Interpretation . . . 74
5.4 Daljan River Delta . . . 75
5.4.1 Drilling and core logging . . . 75
5.4.2 Geophysical investigations . . . 77
5.4.3 Interpretation . . . 77
5.5 Lini . . . 79
5.5.1 Drilling and core logging . . . 79
5.5.2 Geophysical investigations . . . 82
5.5.3 Interpretation . . . 84
5.6 Discussion . . . 84

6 Geomorphology 87
6.1 Morphological markers . . . 94
6.1.1 Triangular facets . . . 94
6.1.2 Wind gap . . . 97
6.2 Scarp profiles . . . 99
6.2.1 East coast . . . 99
6.2.2 West coast . . . 104
6.2.3 North . . . 109
6.3 Discussion . . . 111

7 Synthesis 121

Acknowledgements 125

References 127

A Appendix 141

List of Figures

1.1 Overview map of the Lakes Ohrid and Prespa 4

2.1 Stress regimes and associated fault-slip modes 8
2.2 Stress states illustrated as stress ellipsoids and colour coded stress ratio . . . 9
2.3 Separation process illustrated on the dataset of Galicica II 10
2.4 Setup of the SIR 3000 GPR system 12
2.5 Electric resistivity measurements at the Struga site 13
2.6 Electric resistivity geometries for different arrays 14
2.7 MagSus, ERM and GPR measurements at the Kalista Fault 16
2.8 Spatial distribution of TerraSAR-X and SPOT tiles 18
2.9 ILRIS 3D system from OPTECH Inc. 19
2.10 Scarp profile with denomination of sections 20
2.11 Principle of mountainfront sinuosity index measurements 21

3.1 Tectonic units of the alpine collision zone 24
3.2 Overview of the eastern Mediterranean geodynamic situation 25
3.3 Definitions of the Dinaride-Albanide-Hellenide orogen by different authors . 26
3.4 Structural cross section from the Adrian coast to the Neogene basins 27
3.5 GPS velocities with respect to the Eurasian reference frame 28
3.6 Velocity profile across Albania, Macedonia and Greece 29
3.7 Fault plane solutions in the triangle of Albania, Macedonia and Greece . . . 30
3.8 Location of Dardania during Roman Age 31
3.9 Seismicity in the triangle of Macedonia, Albania and Greece 32
3.10 Lithological map of the Lake Ohrid area 33
3.11 Overview of the Lake Ohrid research area 35
3.12 Panoramic view of the Lini halfgraben 36
3.13 Field photographs of regional geology 37
3.14 Field photographs for syn- and postorogenic development 38
3.15 Watersheds and tributary rivers for the Lakes Ohrid Prespa and Mikri Prespa 39

4.1 Palaeostress sites at Lake Ohrid Basin 44
4.2 Field photographs of kinematic indicators 45
4.3 Stress states for each outcrop 46
4.4 Fold axes and derived stress states 48
4.5 Field photographs of palaeostress sites 49
4.6 Spatial distribution of compressional stresses 50
4.7 Spatial distribution of transtensional stresses 52
4.8 Spatial distribution of E-W trending extensional stresses 54
4.9 Schematic development of the Lake Ohrid Basin 57

5.1 Overview map of sedimentological studies . . . 60
5.2 Maps of the study areas . . . 61
5.3 Interpreted photograph of the Lini Peninsula . . . 62
5.4 Corelog of U03 at Struga . . . 63
5.5 Fossil ostracods samples from various cores . . . 65
5.6 GPR and ERM at Struga . . . 66
5.7 Photographs of the recent Chara-belt . . . 67
5.8 Corelog of DA01 at Velestovo . . . 68
5.9 Magnetic susceptibility and XRF-measurements of core DA01 . . . 69
5.10 Segment 230 - 265 m of GPR line 05 at Velestovo . . . 70
5.11 Sidescan sonar data from Ohrid Bay . . . 71
5.12 Corelog of PL03 at Sveti Naum . . . 73
5.13 GPR line 56 at Sveti Naum . . . 74
5.14 GPR line 32 and ERM at Sveti Naum . . . 75
5.15 Core log of DL01 at Daljan Delta . . . 76
5.16 Fining upward sequence in core DL01 . . . 77
5.17 GPR line 1 at Daljan Delta . . . 78
5.18 Core log of L01 at Lini . . . 80
5.19 Core log of L02 at Lini . . . 81
5.20 GPR line 23 at Lini . . . 82
5.21 GPR line 13 at Lini . . . 83
5.22 ERM at Lini . . . 84

6.1 Overview map of geomorphological investigations . . . 88
6.2 Field photographs of geomorphological features . . . 89
6.3 Schematic postglacial evolution of bedrock fault scarps . . . 90
6.4 Fault displacement vs. erosion/deposition and related scarp evolution . . . 91
6.5 Field photographs of geomorphological features . . . 92
6.6 T-LiDAR image of a fault scarp at Galicica Mountains . . . 93
6.7 Field photograph and sketch of triangular facet evolution . . . 95
6.8 Mountainfront Sinuosity Index for Gorenci and Hudenisht . . . 96
6.9 Wind gap and hanging valleys . . . 98
6.10 Boces fault scarp . . . 99
6.11 Koritsi Rid fault scarp . . . 100
6.12 Lako Signoj fault scarp . . . 101
6.13 Elsani 1 fault scarp . . . 102
6.14 Dolno Konjso fault scarp . . . 103
6.15 Sveti Arhangel fault scarp . . . 105
6.16 Sveti Arhangel photographs . . . 106
6.17 Hudenisht fault scarp . . . 107
6.18 Piskupat fault scarp . . . 108
6.19 Concept of fault scarp evolution at the Piskupat fault . . . 109
6.20 Boces fault scarp . . . 110
6.21 Delogozdi fault scarp . . . 111
6.22 Throw rates at fault scarps by various authors . . . 114
6.23 Sketch of the major neotectonic features at Lake Ohrid . . . 119

A.1 Tangent lineation plots of the east coast outcrops . . . 141
A.2 Tangent lineation plots of the west coast outcrops . . . 144

List of Tables

4.1 Palaeostress sites 42

5.1 Radiocarbon dates of core DA01 69

6.1 Scarp profile parameters 116

1. Introduction

The Ohrid Basin (40°54′ – 41°10′N, 20°38′ – 20°48′E; fig. 1.1) hosts a cross-boundary lake called Lake Ohrid (Macedonian: Охридско Езеро, Ohridsko Ezero; Albanian: Liqeni i Ohrit) which is shared by the Republic of Macedonia (from hereon referred to as Macedonia) and the Republic of Albania (fig. 1.1). In 1979, the lake was declared a World Heritage site by UNESCO because "the city and its historic-cultural region are located in a natural setting of exceptional beauty, while its architecture represents the best preserved and most complete ensemble of ancient urban architecture of the slavic lands" (UNESCO, 2013).

The fresh-water lake is the deepest lake of the Balkans with a maximum depth of almost 290 m. It covers an area of 358 km^2 with an extent of c. 30 km N-S and c. 15 km E-W. Lake Ohrid, the Korca Basin and the sister-lakes Prespa and Mikri Prespa (see fig. 1.1), belong to a group of intramontane basins that originated from back-arc extension during the Late Tertiary in the Dinaric-Albanide Alps. The exact age of the formation of the so called "ancient lake" and its structural context are not well known. There are controversies about the definition of ancient lakes and their minimum age; however, it is commonly accepted that a body of water must have continuously existed since at least the last interglacial period (i.e.~120,000 years ago; SIAL, 2013). Other authors give minimum ages which date back to before the Pleistocene (Gorthner, 1994; Martens, 1997).

Most probably, Lake Ohrid is one of the oldest lakes in Europe with an age of 2 - 5 Ma. Albrecht and Wilke (2009) proposed a minimum age of 2.3 Ma based on molecular systematics and Lindhorst et al. (2012b) assume a minimum age of 1.9 Ma. Ancient lakes have long been recognised as hot spots of biodiversity and the oligotrophic Lake Ohrid preserves an outstanding variety of more than 200 endemic species (e.g. Stankovic, 1960). This makes it the lake with the highest concentration of endemic diversity in the world when considering its small surface area (compare to Lake Baikal with 982 described endemic species and a surface area of 31,722 km^2; Albrecht and Wilke, 2009; Goldman et al., 1996; Martin, 1994; Shimaraev et al., 1994).

Molecular clock analyses of mitochondrial DNA genes from multiple endemic invertebrate species flocks from Lake Ohrid (Albrecht et al., 2006; Trajanovski et al., 2006; Wilke et al., 2007) indicate that there was a significant evolutionary step in early Pleistocene. Until now the reasons for this speciation are poorly understood. It is most likely that major geological and/or environmental events have had an impact on the speciation mechanisms of the endemic faunal elements in Lake Ohrid. It therefore represents a first class site to investigate the impact of geologic, climatic, and environmental events on biological evolution in lakes (Albrecht and Wilke, 2009). Lake Ohrid was included as a target area of the International Continental Scientific Drilling Program (ICDP) in 1993. However, appropriate site survey data were lacking due to accessibility reasons. Therefore no drilling has been carried out by the ICDP to date.

Deep drilling in Lake Ohrid will provide a wealth of scientific information covering a range of disciplines. The main objectives are to drill a continuous archive containing information on climatic conditions, volcanic activities in the central northern Mediterranean, tectonics and mass wasting events since Pliocene. This will provide an insight into the processes which have led to the high degree of endemism within the lake, and also the connection between evolutionary patterns and geologic/climatic events can be determined. Furthermore, information on age and origin of Lake Ohrid will be gained. A workshop for the initiation of the SCOPSCO project (Scientific Collaboration On Past Speciation Conditions in Ohrid) was held in autumn 2008; the drilling of the deep holes was carried out from April to June 2013 by ICDP.

In the last few years several groups have been working in the area to address some of the key topics of the SCOPSCO project. Lindhorst et al. (2010) were investigating the sediment architecture of the lake basin using seismic data to image the depositional patterns of the past and sidescan sonar to map the recent sedimentary patterns and the interplay between mass movements and tectonic lineaments as well as the areal extent of bottom morphology patterns. Vogel et al. (2010a,b) were working on the upper sediment succession of offshore cores to reconstruct the sedimentary history of Lake Ohrid on different time scales. They were able to provide a chronological framework for Lake Ohrid by correlating explosive eruptions of Italian volcanoes with volcanic tephra layers. Sediments were dated back to Marine Isotope Stage 6 (136 ka). Research also has been carried out on the origin of invertebrates (diatoms, oligochaetes, leeches, poriferans, tricladids, molluscs, ostracodes) and fish species (Albrecht and Wilke, 2009; Hauffe et al., 2011; Kostoski et al., 2011; Wagner and Wilke, 2011). This was done to provide information on the lakes limnological history and on the timing of major evolutionary events leading to the extraordinary diversity and endemism. These molecular clock approaches provide a minimum age of the lakes and give insight into the evolution of the watersheds of Lakes Ohrid and Prespa, e.g. by dating the separation of sister taxa.

The work presented in this thesis focused on the tectonic framework and the geomorphological evolution of the Lake Ohrid Basin. A multidisciplinary approach was chosen to meet our aim. Inferences from field observations are compared with geophysical and geodetic data to assess rates and mechanisms of present-day processes. As Lake Ohrid is an ideal location to analyse the interactions between sedimentation and active tectonics, the objectives were defined as follows:

Old structures:

- Determine of the palaeostress fields which controlled the evolution of the Lake Ohrid Basin by structural and palaeostress analysis.

Recent Structures:

- Gain information on fault orientation, fault geometry, spatial distribution of fault scarps across the basin, the influence of lithology on faulting mechanisms, and slip rates.

- Gain insight into the grade of activity in the Lake Ohrid Basin.
- Determine if identified tectonic structures are capable of causing major speciation events and highlight possible triggering mechanisms.

Landscape Evolution:

- Analyse the role that alluvial fans and delta formation have had on landscape evolution around Lake Ohrid.
- Identify areas experiencing tectonic deformation. Define active/less active areas throughout the basin.

1.1. Scope of the thesis

The Lake Ohrid area has undergone several phases of deformation since Tertiary (see chapter 3.2). Even today it forms one of the most active seismic zones in Albania/Macedonia with several moderate earthquakes reported during the last few centuries (Muco, 1998; NEIC, 2013) and a few major earthquakes during historical times (Reicherter et al., 2011). The landscape of the Ohrid area exhibits a variety of features that are connected to this seismic activity within the basin. This study provides insight into the processes that the Ohrid Basin was exposed to since the Quaternary and the derived consequences for landscape formation. As neotectonic surveys need input from a variety of geoscientific disciplines and archaeology, e.g. sedimentology, tectonic morphology, dating methods, palaeoseismology, archaeoseismology, geodesy, remote sensing, geophysics and structural geology (for further reading see Bull, 2007; Burbank and Anderson, 2001; McCalpin, 2009; Stewart and Hancock, 1994), a multidisciplinary approach was necessary for the success of the study. Inferences from field observations are compared with geophysical and geodetic data to assess rates and mechanisms of present-day processes.

The study covers a wide range of topics beginning with an initial summary of the study area‘s regional setting, its geological structures, lithology and particularities. Then, suitable methods and method combinations are introduced for a neotectonic study such as this, as the quality of methods and the right choice of material has a major impact on the success of a field campaign. Palaeostress and structural analyses were undertaken to get an idea of the past and recent tectonic stress regimes; this gives an impression of what the geomorphology should look like and provides data on the current deformational system. Detailed sedimentary analysis using a variety of methods were carried out to distinguish several coastal domains and sedimentary realms. This also provides a lot of information on how the lake responds to landscape change and how the lake level varied during the Holocene. A detailed geomorphological study was undertaken which deals with the expression and geometrical properties of fault scarps around Lake Ohrid. This includes analysis of scarp lengths and spatial distribution and a survey on how lithology interacts with scarp formation around lake Ohrid resulting in insight into localised basin activity. To conclude, the following question is answered: is the seismic activity together with the geological character of Lake Ohrid capable of triggering speciation events in the lake?

To summarise: Lake Ohrid's setting in an extensional back arc environment, the extraordinary landscape morphology, the age of the lake together with its continuity since the Pliocene, and the presence of a high endemic diversity makes the lake a valuable archive for studying landscape forming and its effect on species evolution.

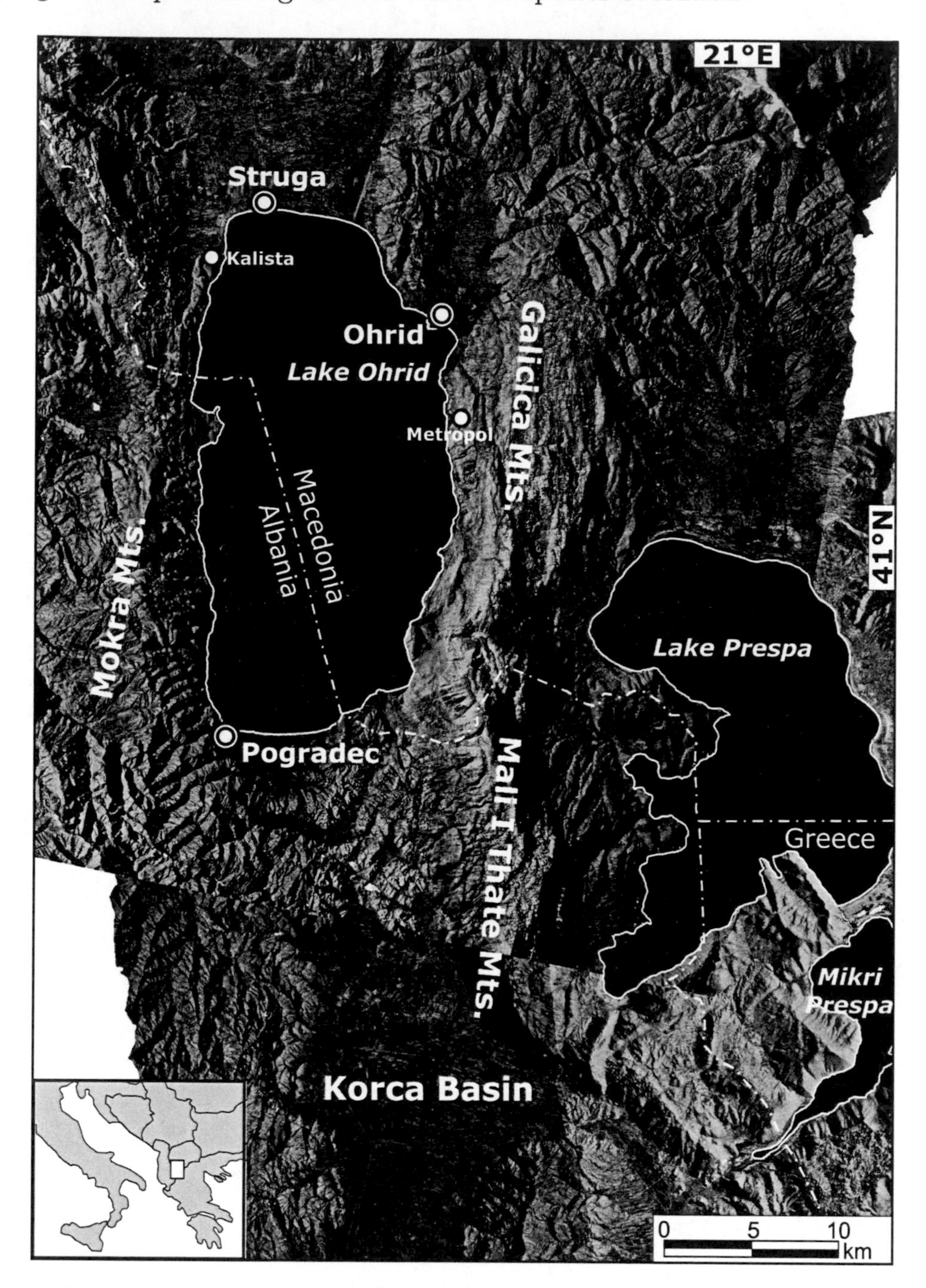

Figure 1.1.: Overview map of the Lakes Ohrid and Prespa using TerraSAR-X satellite images (DLR, 2013). Due to different looking angles, the tiles distribute different shadows.

1.2. Structure of the thesis

The text of this thesis is mainly based on the following papers:

Chapter 2:
Reicherter, K., Hoffmann, N., Lindhorst, K., Krastel, S., Fernández-Steeger, T., Grützner, C. and Wiatr, T. (2011): Active basins and neotectonics: morphotectonics of the Lake Ohrid Basin (FYROM and Albania). Zeitschrift der deutschen Gesellschaft für Geowissenschaften, 162: 217-234, Stuttgart.
Chapter 2 represents a summary of all methods used during the field campaigns, in the laboratory and in further analyses. A combination of these methods using a multidisciplinary approach was successfully carried out within the Lake Ohrid Basin and evidence of recent faulting was identified.

Chapter 3:
Hoffmann, N., Reicherter, K., Fernández-Steeger, T. and Grützner, C. (2010): Evolution of ancient Lake Ohrid: a tectonic perspective. Biogeosciences, 7: 3377-3386.
This chapter focuses on the tectonic evolution of the Lake Ohrid area and describes tectonic features that are present in the basin and surrounding areas. The geological and geodynamical settings are summarised and the main units are discussed.

Chapter 4:
Hoffmann, N., Reicherter, K., A., Arndt, M. (in prep.): Palaeostress History of the Lake Ohrid Basin (Albania/FYR of Macedonia).
The main objective of this chapter was to analyse the structures affected by brittle deformation, to provide a (neo-) tectonic history of the lake and to show how the basin initially evolved. Additionally, the driving forces are highlighted which have affected the area since Miocene.

Chapter 5:
Hoffmann, N., Reicherter, K., Grützner, C., Hürtgen, J. Rudersdorf, A., Viehberg, F. and Wessels, M. (2012): Quaternary coastline evolution of Lake Ohrid (Macedonia/Albania). Central European Journal of Geosciences. 4(1): 94-110.
In this chapter the lake-shore and deltaic deposits of the major inflows were studied in order to obtain information on Holocene lake level changes and tectonically induced coastline modifications which are most likely in this seismic active area (see Hoffmann et al., 2010; Reicherter et al., 2011). The obtained sediment cores and shallow geophysics (ground penetrating radar and electric resistivity) revealed the large-scale sediment architecture. Grain size and geochemical analyses as well as micropalaeontological investigations supplement the results.

Chapter 6:
Hoffmann, N. (submitted 2012): The seismic landscape of Lake Ohrid (FYROM/Albania). Annals of Geophysics
The geological and geodynamic setting of Lake Ohrid results in a diverse seismic landscape that provides tectonically induced landscape features which are still preserved in the surroundings of Lake Ohrid. In this chapter a link between landscape evolution and tectonic activity in the region is provided. Tectonic geomorphology is used to classify the seismic activity of an area. Integrating different approaches of this discipline with

high resolution satellite images and field work can gain a detailed picture of the landscape response to neotectonic movements.

Chapter 7:
Synthesis
Summary of the conclusions of the preceding chapters and their relevance for the objectives.

2. Methods

In order to assess the active tectonic features and seismicity of the region, the implementation of a combination of different geoscientific methods is necessary. “Classical” tectonic techniques like field mapping, structural and palaeostress analysis, and evaluation of the historical and instrumental seismicity of the area accompanied by shallow geophysics (ground penetrating radar, resistivity) and remote sensing methods (satellite images, digital elevation models, terrestrial laser scanning) are required to fully understand the active tectonics within the region.

The investigation area is limited to the Ohrid Basin itself. It is bounded by the crests of the Galicica/Mali i Thate and Mokra Mountains to the east and west, to the north by the outlet of the Crni Drim and to the south by the drainage divide between Ohrid and Korca basins.

2.1. Palaeostress Analysis

Anderson (1942) was one of the first that related stress fields to fault kinematics. But he considered only newly formed, pure dip-slip or pure strike-slip faults (see fig. 2.1), and this is why these type of faults are termed Andersonian. Oblique faulting or reactivation along existing weakness zones was not considered in his study; therefore these fault types are defined as non-Andersonian. The analysis of palaeostress data is based on the principle that striations on a fault plane represent the displacement direction of the hanging wall block in relation to the footwall block. The striations can therefore be correlated with the palaeostress tensor and the stress conditions that caused the brittle structures under consideration.

The stress field or state of stress is defined by the orientation and magnitude of the three mutually perpendicular principal stress axes σ_1, σ_2 and σ_3 and describes the force per unit area depending on the applied loads on a body.

The study of the evolution of the palaeostress field in the Lake Ohrid Basin is mainly based on the collection of structural data such as fault-slip data but also folds, joints and fractures. Palaeostress data were collected in the vicinity of the lake, where Palaeozoic and Mesozoic rocks crop out. This provided lithologies with favourable conservation conditions for kinematic indicators (figs. 6.5 and 4.2) such as limestones, harzburgites and serpentinites. Depending on their age, these rocks bear the imprints of several deformation phases that have affected the basin system from the Late Cretaceous to present. At each location a representative number of fault slip data were measured on a fault plane (minimum five values) recording the spatial orientation of the fault plane (dip direction, dip) and striae (azimuth, plunge), and also the sense of slip (reverse, normal, dextral or sinistral). The sense of slip was determined by the identification of different kinematic indicators such as crystal fibres, striations, steps, riedel shears, etc. as summarised by Doblas (1998). Furthermore, the quality of data and rock types were also assigned to each dataset (see also table 4.1).

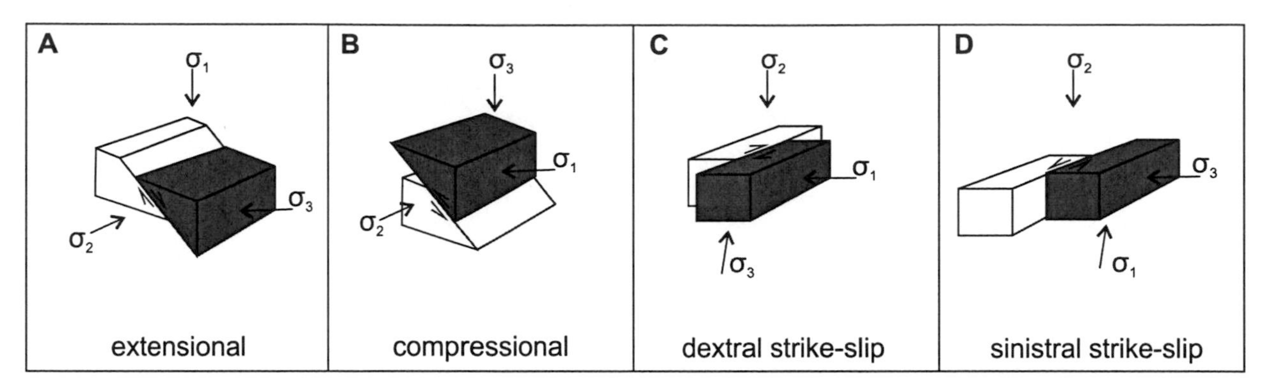

Figure 2.1.: Stress regimes and associated fault-slip modes according to Anderson (1942). A newly formed fault plane is oriented parallel to σ_2. The sense of shear along the faults is normal (A), reverse (B), or strike-slip (C) and (D), depending on which of the principal stress axes, σ_1, σ_2, or σ_3, respectively, is vertical. The dip angle between the fault plane and the σ_1-axis depends on material properties. Olique stress regimes are considered as non-Andersonian with none of the principal axes being vertical representing, therefore, a mixture of the above mentioned fault-sip modes.

According to the Wallace and Bott hypothesis (Angelier, 1994; Bott, 1959; Wallace, 1951) the slip on a fault plane (documented by the orientation of striation) is parallel to the maximum resolved shear stress $\tau_{\max}$ on this very fault plane. It has therefore a linear correlation with the stress tensor σ that activates a fault with the normal of the fault being n (see Yamaji, 2007, for details).

$$\tau_{\vec{\max}} = \sigma \cdot \vec{n} - (\vec{n} \cdot \sigma \cdot \vec{n}) \cdot \vec{n}$$

Considering the Mohr-Coulomb failure criterion, a fault plane, with an orientation that matches the applied stress, will form when the shear stress overcomes the cohesive strength and internal friction of the material. This is expressed by the following equation where σ_{N} = normal force on the shear plane; C = cohesion specific to the material under consideration; and μ = internal friction coefficient:

$$\tau = C + \mu \cdot \sigma_{\mathrm{N}}$$

This principle allows the calculation of the stress tensor by inversion of the fault-slip field data. Several methods have been proposed to calculate stress states from field data (Angelier, 1990, 1994; Ramsay and Lisle, 2000; Reches, 1987; Sippel et al., 2009, 2010) based on either slip criteria, frictional criteria or a combination of both.

Here the Stress Inversion Via Simulation Method (SVS) of Sippel (2008) was applied. This technique combines the PBT-axes method of Sperner et al. (1993) with the Multiple Inverse Method (MIM) of Yamaji (2000). The datasets were analysed using commercial software: TectonicsFP by Reiter and Acs (2012) and MIM by Yamaji (2000). The results are obtained in terms of a reduced stress tensor, consisting of: (1) orientations of the three principal stresses σ_1, σ_2 and σ_3 with $\sigma_1 \geq \sigma_2 \geq \sigma_3$; and (2) the ratio of principal stress differences Φ (see fig. 2.2; Angelier, 1994):

$$\Phi = (\sigma_2 - \sigma_3)/(\sigma_1 - \sigma_3) \text{ with } 1 \geq \Phi \geq 0$$

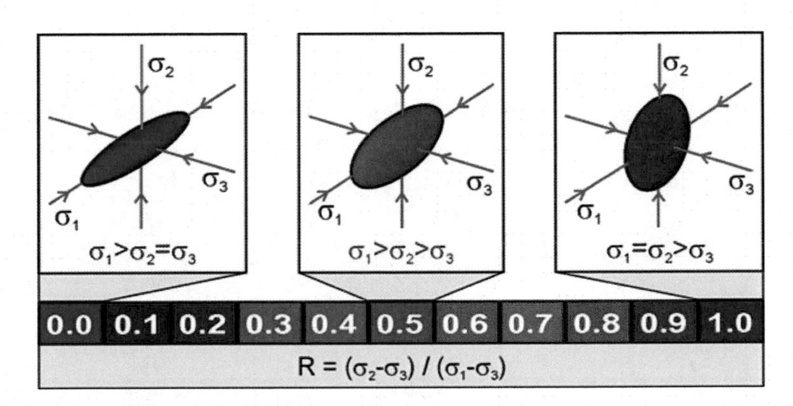

Figure 2.2.: Different stress states illustrated as stress ellipsoids which are spanned by the three stress axes, $\sigma_1 \geq \sigma_2 \geq \sigma_3$. different values of stress ratio Φ are displayed in a colour code which is used throughout the study. $\Phi = 0.0$ indicates the magnitudes of σ_2 and σ_3 to be equal; $\Phi = 1.0$ expressing σ_1 and σ_2 to be equal in magnitude. From Sippel (2008).

The SVS requires four steps of data processing (see also fig. 2.3):

(1) The process of field data correction, which rotates the striae to align on the fault plane, PBT-axes are calculated for every fault-slip datum with the P-axis in the direction of compression, the T-axis in the direction of extension and the B-axis as a neutral axis that lies along the strike of the fault plane (fig. 2.3A).

Fracture angles experimentally obtained by (Byerlee, 1968; Hubbert, 1951; Jaeger and Cook, 1979) range between 20° and 40°. Here a fracture angle of 30° was assigned as this angle has been proven to work for fault slip data obtained from the field (Reicherter and Peters, 2005; Sperner, 1996). From the plot of PBT-axes, fault clusters, representing kinematically homogeneous subsets, are identified and separated (fig. 2.3B). Data that could not be assigned to any of these subsets are treated as remnants. These mainly comprise non-horizontal or non-vertical faults which are defined as oblique or non-Andersonian faults (Anderson, 1942; Sippel, 2008).

(2) The reduced stress tensor is then calculated for each subset using both the Numeric Dynamic Analysis (NDA; Spang, 1972) and the Multiple Inverse Method (MIM; Yamaji, 2000), which both revealed very good results (Sippel et al., 2009, fig. 2.3C).

(3) The next step is to simulate the stress state (fig. 2.3D). Here it is important to obtain low misfit angles (β), which are the angular difference between the calculated maximum shear stress and the observed slip direction along a fault plane. Principal stresses needed for input are derived from the mean vectors of the associated PBT-axes, whereas the stress ratio value Φ results from the run through the MIM. For further details see Sippel (2008).

(4) Finally, a second simulation provided an improvement of data. Within this simulation the obtained stress states from step (3) are tested against the entire fault population to estimate its potential to activate other faults from the set. A fault slip datum with a misfit angle $\beta \leq 30°$ was regarded as being activated by the stress field in this test. So that the fault population of each site is qualitatively divided into different subsets, each being consistent with its specific stress regime. The results of the second simulation are summarised in a stress state plot with information on the three principal stresses and the stress ratio (fig. 2.3E)

The distribution of regional palaeostress patterns and their evolution over time needs the input of data from multiple faults at various locations. To work out the relative chronology of stress regimes, different sets of intersecting striations and fold-axes data were used.

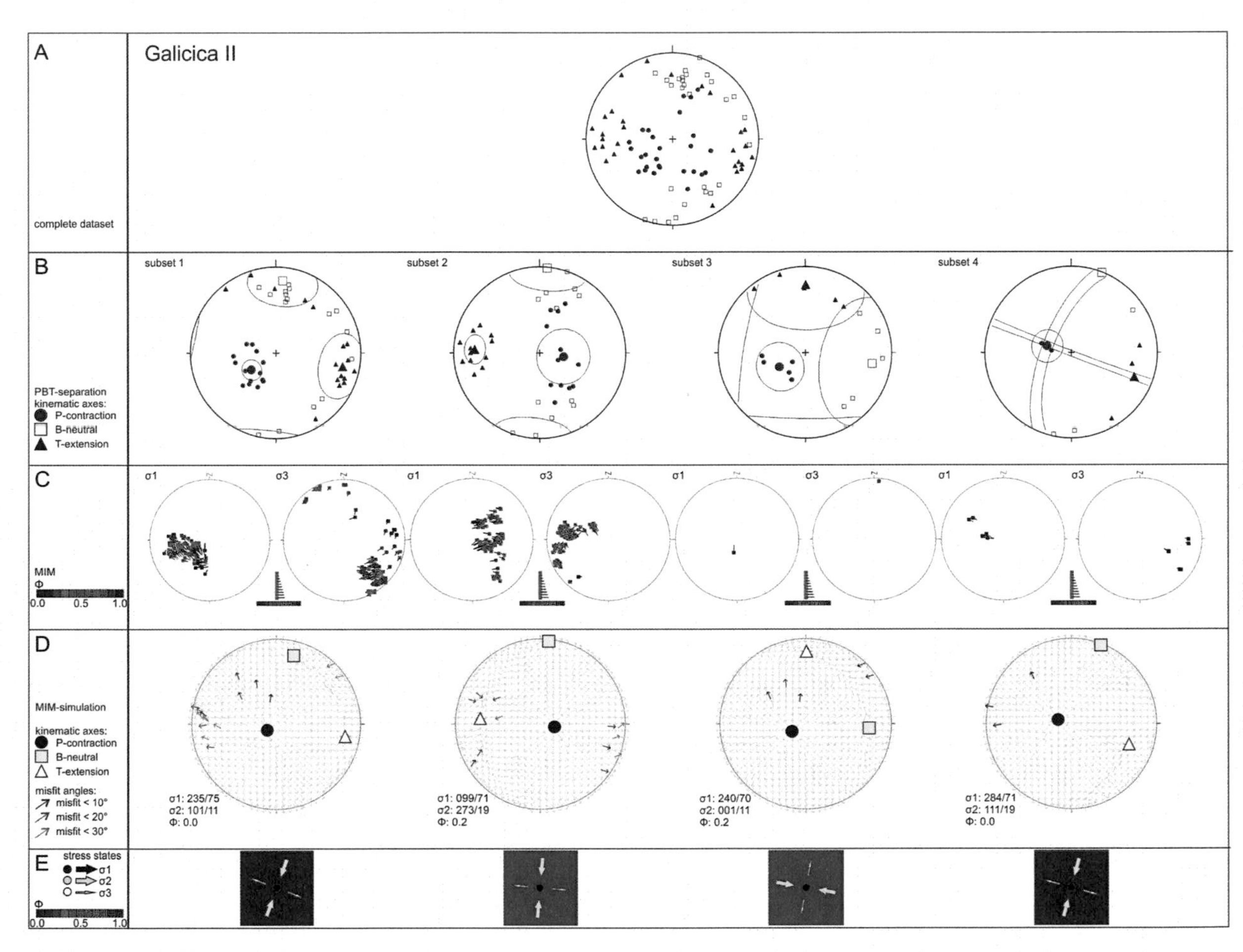

Figure 2.3.: Separation process illustrated on the dataset of Galicica II. A: PBT-axes plot of the complete dataset. B: Separation of the dataset into subsets. C: Running individual subsets through MIM. Clusters show homogeneous data. D: Tangent lineation plot showing results of stress simulation in MIM. Misfit angles are displayed as colour code of the arrows. Kinematic axes were initially derived from PBT-subsets and adjusted to find the best fit. E: final stress states showing the stress ratio Φ and the three principal stresses.

2.2. Geophysical Techniques

Geophysical techniques, especially shallow geophysics, can be used to quickly gain data from a wide area and to assess the study region for subsequent investigations. A variety of non-destructive methods can be supplemented by shallow drilling operations for a precise and prompt analysis of the sedimentary sequence, for ground truthing, calibration of the data, for sampling and further laboratory analyses.

2.2.1. Ground Penetrating Radar

Ground penetrating radar (GPR) is a non-invasive geophysical method for subsurface investigations which is applied widely in various geological settings. It is based on propagating electromagnetic waves which are emitted underground and reflected off boundaries between different materials with varying electromagnetic properties. The propagation of electromagnetic waves in solid media is mostly influenced by dielectric permittivity and conductivity. The dielectric permittivity controls the amplitude of reflections and the velocity of the electromagnetic wave. Changes in the dielectric contrast at layer boundaries are determined by variations in physical properties such as water content, salinity, porosity, grain size, or mineral composition, that produce reflection waves which can be registered and analysed. The conductivity controls the attenuation of the waves. High conductivity leads to high attenuation and low penetration depths (Neal, 2004).

GPR was applied at all drilling locations to gain a detailed picture of the subsurface and the lateral extent of identified features. With GPR wide areas an be covered with low effort and good results.

For the survey, a SIR 3000 GPR system by GSSI with a survey wheel, GPS, and two antennas (270 MHz and 100 MHz) was used to cover different resolutions and depths (fig. 2.4). The 270 MHz antenna covers a penetration depth of 3 - 7 m with a maximum vertical resolution of 7 cm. The 100 MHz antenna reaches penetration depths of maximum 30 m and has a resolution in the decimeter - metre range. For analysis it turned out that the 270 MHz antenna is the more sufficient at least concerning the Lake Ohrid area; as the resolution of the 100 MHz antenna was not good enough for our purposes and to filter out even small reflectors or follow small structures. In the lower areas of the plains the high water table lets the signal diminish at above 7 m and in higher altitudes the soil cover was so scarce that the bedrock was the limitation factor for deeper penetration depths. In general, penetration depths deeper than 10 m were not reached.

The data have been processed with the ReflexW software (version 4.5.5 by Sandmeier, 2010). The main processing steps accomplished were: move start-time, background removal, and gain adjustment. A topographic correction was not carried out as the elevation differences in the field were marginal. A butterworth bandpass (frequency filter) was applied with half the antenna frequency as the lower threshold and twice the nominal frequency as the upper threshold. A depth axis was assigned to the radargrams on the base of an electromagnetic wave velocity of 0.1 m/ns (Neal, 2004) which is an estimated value for depth evaluation. However, velocity determination by hyperbola analysis led to the same value. GPR lines were taken parallel and perpendicular to each other in order to systematically image the subsurface.

Figure 2.4.: Setup of the SIR 3000 GPR system using the 270 MHz antenna. Photograph by Sandra Fuhrmann.

2.2.2. Electrical Resistivity

Electrical resistivity measurements (ERM) are widely used in geotechnical, geological and archaeological investigations. Similar to other geophysical techniques, ERM is non-invasive and results are relatively fast to attain (fig. 2.5). The method is based on measuring the distribution of specific electric resistivities in the subsurface. A known electrical direct current (DC) is injected into the ground via two electrodes and the difference of potential (or voltage) is measured by two other electrodes (4-point ERM method). The apparent specific resistivity of a single point is then determined using Ohm's law. As mineral grains solely are not conductive (except some materials such as metallic ores), the resistivity of soils and rocks is mainly controlled by the pore water conditions (amount of pore water, salinity, pore space, etc.).

Moving the electrodes along a profile and increasing their distance and spacing allows recording the values for different depths and locations. From these so-called "pseudosections", the model of the resistivity distribution is then inverted using specific software. The investigation depth, resolution and the measurement sensitivity vary with different geometries and configurations (e.g., Wenner array, fig. 2.6A; Schlumberger fig. 2.6B; dipole-dipole array, fig. 2.6C). The Wenner array is suitable to detect steeply dipping features with strong resistivity contrasts, whereas the Schlumberger method is mainly used to determine depth and resistivity values for horizontal structures. Dipole-dipole arrays are useful for measuring lateral resistivity changes.

Figure 2.5.: Electric resistivity measurements at the Struga site. Photograph by Sandra Fuhrmann.

DC-resistivity measurements with the “4-Punkt light” ERM system from Lippmann Geophysikalische Messgeräte (fig. 2.5) were applied; inversions were computed with RES2DINV v. 3.58 by Geotomo software. Only three inversion steps were calculated but since the errors were relatively small, they did not affect the interpretation. Resistivity profiling was carried out parallel to a number of the GPR survey lines. Further profiling was applied along strike and perpendicular to fault zones to detect lateral variations. For all profiles Wenner, Schlumberger and dipole-dipole arrays were measured to see what works best and to work out different features. In this thesis only the profiles are shown that gave the best image. It turned out that the Wenner array delivered the best results in this environment and for our purpose.

2.2.3. Sidescan Sonar

A Klein 3000 dual frequency Sidescan-sonar was used by the SCOPSCO team from a boat with 100 kHz and 500 kHz transducers connected to a lightweight cable for investigations of the lake floor morphology. The entire area of Ohrid Bay was mapped up to 100 m water depth, with profiles covering widths of 150 or 200 m. Data were then processed with the mosaicking software SonarWiz. A georeferenced image with 0.5 m resolution was used for further analysis.

2.3. Percussion drilling

Percussion drilling was used to get information on the sedimentary sequence and lithology at certain points of interest and additionally to verify the position of the ground water table detected by GPR and/or ERM. The cores provide insight into sedimentary processes and into the evolution of the plains around Lake Ohrid. Drilling was carried out with a Cobra hammer and extension rods with a maximum total length of 10 m. At all the drilling

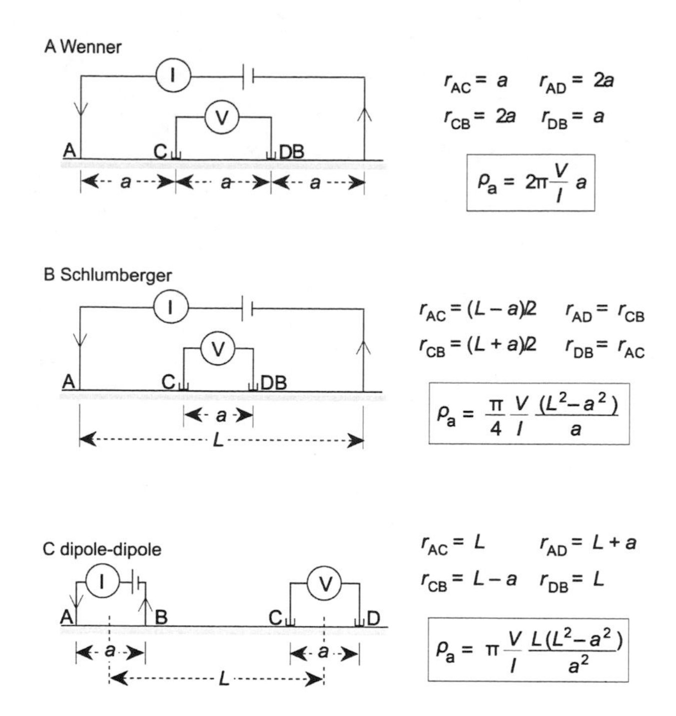

Figure 2.6.: Electric resistivity geometries of current and potential electrodes for A: Wenner, B: Schlumberger and C: dipole-dipole configurations. Modified after Lowrie (2007).

sites, an open-window sampler was used and the cores were described directly in the field. Additionally, at specific sites liner cores of max. 9 m depth were taken for subsequent laboratory analysis. In the laboratory the cores were sliced in half, photographed, and the magnetic susceptibility was measured. Samples were washed through a set of sieves with mesh sizes of 0.063 mm, 0.125 mm, 0.250 mm, 0.500 mm, 1.000 mm, 2.000 mm and 4.000 mm to filter out shells, wood chips and other organics as well as microfossils. The different fractions were dried and screened under a microscope for organic material and microfossil remains. Microfossils are a criterion for the depositional milieu and transport mechanisms, organic material was used for ^{14}C dating. The ostracods were identified by the works of Klie, Mikulic, Jurine and Petkovski (Jurine, 1820; Klie, 1939a,b, 1942; Mikulić, 1961; Petkovski, 1959, 1960a,b,c, 1969a,b; Petkovski and Keyser, 1992). The Munsell Rock Color Chart was used both in the field and in the lab to determine the soil color. Some cores were scanned using X-ray fluorescence (XRF) in order to obtain their chemical properties.

2.3.1. X-ray fluorescence analysis

XRF analysis is a non-destructive method to identify element concentrations in a rock sample or sediment cores with a high resolution. This technique was applied to the DA01 core (see chapter 5 for details) to obtain information on deposition processes and environment and to possibly identify tephras or cryptotephras which can be compared to data from Vogel et al. (2010a,b). Unfortunately, the found cryptotephra could not be linked to the tephrochronology of Vogel et al. (2010a). The analysis was carried out at the Institute of Geology and Mineralogy at the University of Cologne using an ITRAX core scanner (COX Ltd.). The scanner is equipped with a Mo-tube set to 30 kV and 30 mA and a Si-drift chamber detector (for details see Vogel et al., 2010b). This scanner gives out continuous high resolution element concentrations in the range from Si to U (Rothwell and Rack, 2006). Scanning resolution was set to 1 mm with an analysis time of 10 s per measurement. The relative concentrations of the elements were derived as an equivalent to count rates.

2.3.2. Magnetic Susceptibility

Magnetic susceptibility (MS) is widely used to distinguish materials from different origins (e.g., Da Silva et al., 2009; Ellwood et al., 2000; Mullins, 1977) depending on their mineral content. MS is the ability of a rock to become temporarily magnetised and reflects mineral properties in an external magnetic field. This results in dimensionless SI units as MS is calculated by the magnetisation of the rock sample divided by the magnetic field strength. Ferromagnetic minerals (like hematite and magnetite) and ferruginous minerals have high positive values of MS, whereas diamagnetic minerals like quartz show very weak negative values. Weak, positive values are related to paramagnetic minerals like dolomite and smectite. Sediments generally show differences in their content of such minerals. MS measurements provide additional quantitative information on the sediment composition of the section analysed. The Bartington MS2 system with the MS2K sensor was used for laboratory work on the sediment cores and on a fault gouge in the field. The response volume for the MS2K of the sample is in the order of 0.2 cm^3. The sample interval was between 1 cm and 2.5 cm.

2.3.3. Combining geophysical methods across faults

A combination of the aforementioned techniques was used to verify and accentuate the results. GPR and ERM were used in combination with MS in order to characterise areas, where using only one technique would have been insufficient. A good example of how the different methods deliver the same interpretation is shown in figure 2.7. The N-S trending fault (41°8′56.418"N; 20°38′14.251"E) close to the town of Kalista (fig. 1.1) is only exposed in an abandoned quarry. Besides the outcrop in the quarry no further surface expression of this fault was found. Satellite images, however, show a valley within N-S trending mountainous ridges which may be regarded as evidence for tectonic faulting. In the exposed fault reddish thin-bedded Triassic limestones are displaced against yellowish Jurassic limestones. The rock units are separated by a fault gouge formed of intensely deformed rocks, where leaching has taken place (fig. 2.7A). Magnetic susceptibility measurements across the fault shows major increase in SI value in the fault gouge compared to the surrounding limestone (fig. 2.7B). The fault zone is heavily altered and shows an enrichment of ferruginous minerals like hematite. The GPR profile (270 MHz antenna) across the fault zone, which has been carried out in some ten meters distance to the outcrop, clearly reflects the rugged zone of the fault within the limestones. The radar signal almost diminishes in the fault zone (fig. 2.7C), pointing to higher porosity and water content. Finally, the electrical resistivity section (Wenner array) shows a zone of high resistivity in the hanging wall of the fault and a change into less conductive rocks in the footwall (fig. 2.7D). This combination of shallow, high-resolution geophysical methods allowed to recognise and follow the fault zone, even when no tectonomorphological expressions are present in a distance of the outcrop.

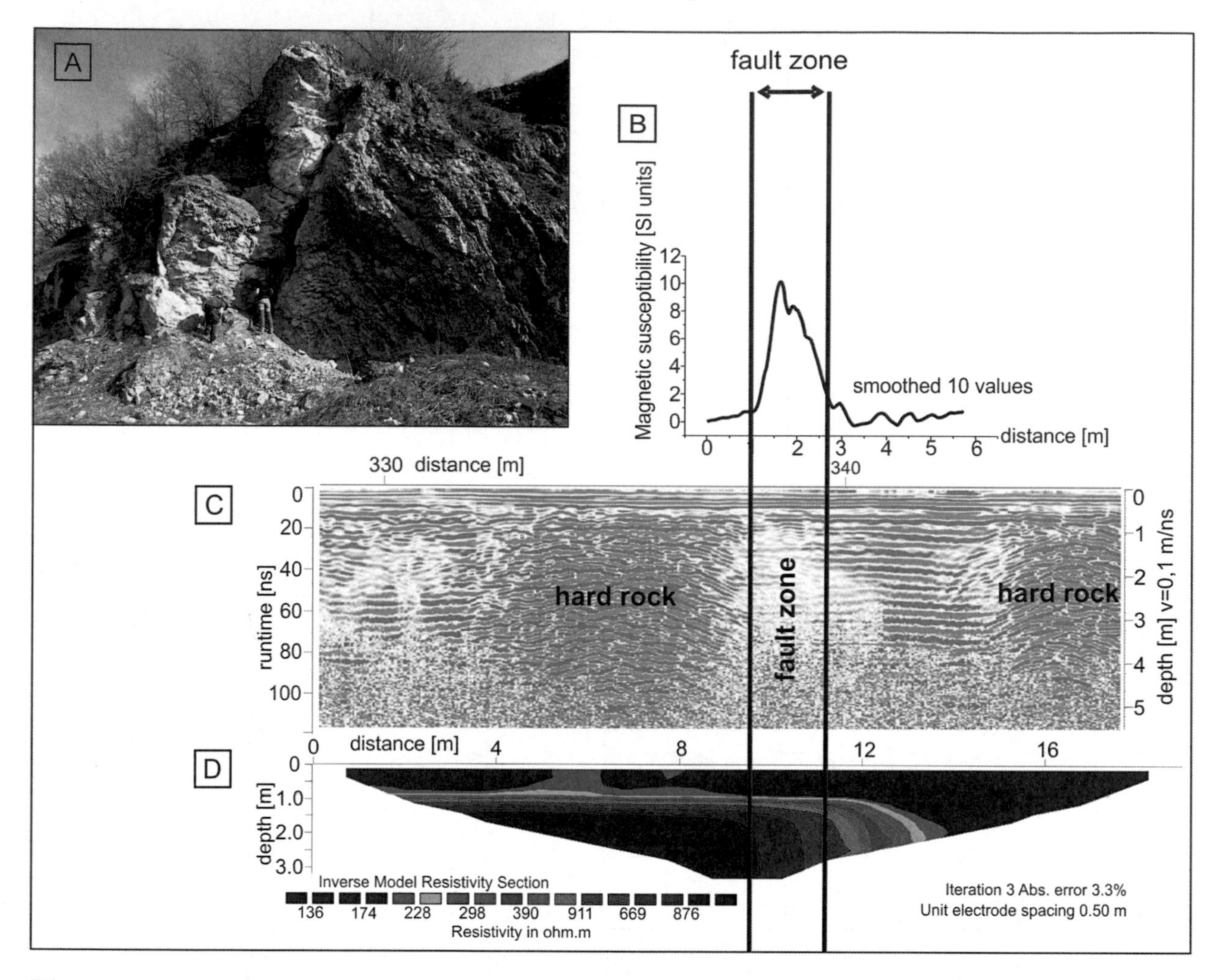

Figure 2.7.: A: N-S trending Kalista Fault outcropping in abandoned quarry (for location see figure 1.1). B: Magnetic susceptibility shows major increase in SI value in the fault gouge. C: GPR profile across the fault zone. D: Electrical resistivity section (Wenner array) shows a zone of high resistivity in the hanging wall of the fault and a gradual change into less conductive rocks in the footwall. From Reicherter et al. (2011).

2.4. Dating methods

2.4.1. Radiocarbon dating

Radiocarbon dates from organic material such as leaves and roots were measured at the Keck Carbon Cycle AMS Laboratory of the Earth System Science department of the UC Irvine. Three samples of the DA01 core (see chapter 5 for details) were analysed. Radiocarbon concentrations are given as fractions of the modern standard, $d^{14}C$, and also as conventional radiocarbon age following Stuiver and Polach (1977). Sample preparation backgrounds have been subtracted based on measurements of ^{14}C-free wood (organics) and calcite (carbonates). All results have been corrected for isotopic fractionation with $d^{13}C$ values measured on prepared graphite using the AMS spectrometer. $D^{13}C$ values were measured to a precision of $< 0.1‰$ relative to PDB standard (Pee Dee Belemnite used as international reference standard for stable isotope ratios of carbon), using a Thermo Finnigan Delta Plus stable isotope ratio mass spectrometer (IRMS) with Gas Bench input. Dates were finally calibrated into calendar years before present (cal. a BP) using the CalPal2007HULU calibration curve (Danzeglocke et al., 2013)

2.4.2. OSL dating

Samples from a dragged palaeosol at the Metropol location (figs. 1.1 and 3.14C) were used for optical stimulated luminescence (OSL) dating with the intention to get a minimum age of the latest activity along that fault. The horizontal deposited soil was deformed and dragged by fault movement. Four samples were recovered in iron cylinders and analysed by the Department of Geography at the University of Cologne. For the determination of the dose rate radionuclide concentrations were measured with high resolution gamma spectronomy. Equivalent dose determination and age calculation was performed with optical stimulated luminescence on quartz (fine sand fraction 100-200 μm), with the "Single-Aliquot Regenerative Dose Protocol" (after Murray and Wintle, 2000, 2003; Wintle and Murray, 2006) and a aliquot diameter of 2 mm (approximately 100-500 grains per aliquot; after Duller, 2008).

2.5. Tectonic morphology studies

An initial overview of the Lake Ohrid Basin reveals several typical geomorphic indicators which testify to an active, seismic landscape (as defined by Dramis and Blumetti, 2005; Michetti et al., 1995; Michetti and Hancock, 1997; Michetti et al., 2005; Serva, 1995). The most prominent features around Lake Ohrid are well pronounced hard rock fault scarps which have developed in all lithologies and stratigraphic units. Dramis and Blumetti (2005) state that over a long period of active deformation a seismic landscape evolves. This includes not only the development of bedrock fault scarps but also the distribution of a variety of other morphological features such as wine-glass shaped valleys and triangular facets. These characteristics can be quantified and analysed by a variety of techniques which are discussed in the following sections. Geomorphological methods used include construction of scarp profiles, mapping of morphological features with regard to their distribution and shape and the analysis of slip rates derived from field data.

2.5.1. Remote Sensing

We used a combination of available satellite images such as Landsat (30 m spatial resolution from NASA, 2013), Aster (ASTGTM 30 m spatial resolution from USGS, 2012) and SRTM data (3 arc seconds/90 m from CGIAR, 2013) complemented by SPOT and TerraSAR-X data in order to delineate geomorphological features, to map structures and to quantify faults. In addition Terrestrial Laser Scanning (LiDAR) was applied on certain structures. SPOT data were ordered for the Galicica mountain area around Trpejca to obtain more information on the wind gap located between Ohrid and Prespa Lakes, the hanging valleys and fault systems above Trpejca. The spatial distribution is seen in figure 2.8.

TerraSAR-X data are provided by DLR. TerraSAR-X is a German satellite equipped with SAR (Synthetic Aperture Radar) in the X-band and has been operating since 2008. It is producing high-quality radar images of the earth‘s surface from a near-polar orbit at an altitude of 514 km. Data presented here are in "stripmap" mode which cover a 30 km wide strip with a resolution between 2 and 3 m (configurations: ground range; azimuth resolution 3 m, single mode; for further information see DLR, 2013). The high resolution of the tiles allows to identify morphological features and fault traces and measure lengths of distinct features. The spatial distribution of ordered data can bee seen in figure 2.8.

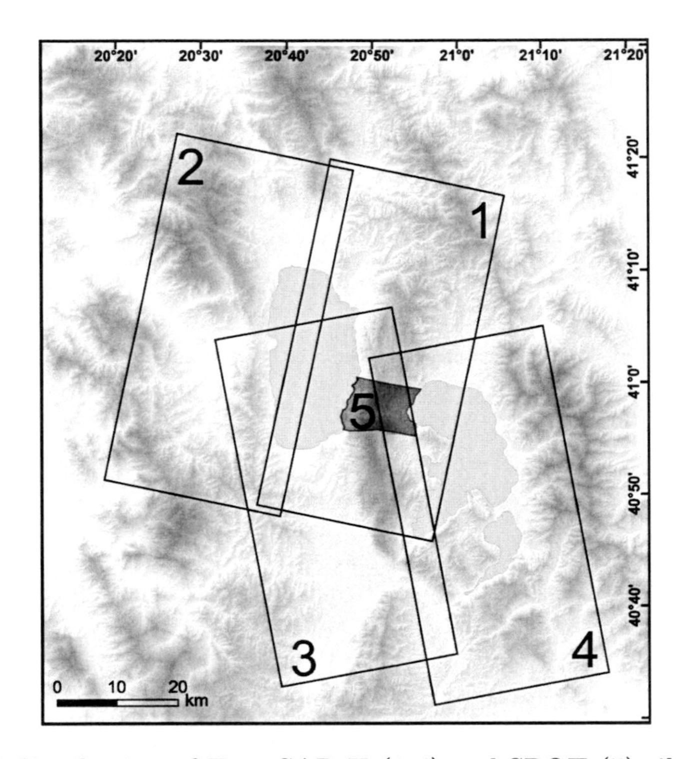

Figure 2.8.: Spatial distribution of TerraSAR-X (1-4) and SPOT (5) tiles.

LiDAR

The morphology and roughness of hard rock fault scarps (e.g. limestones or ophiolites) was observed in high-resolution T-LiDAR images. Other parameters such as the dip of the fault, corrugations and lineations, and the scarp height were determined with laser-scanning observations (fig. 2.9). Ground-based or terrestrial LiDAR (Light Detection and Ranging, T-LiDAR) systems have the capability to detect millimetre-scale asperities and corrugations on a plane and also to create high-resolution digital elevation models. LiDAR is an emergent discipline; however, the method has been established as a versatile data acquisition tool in photogrammetry, engineering technologies, atmospheric studies and many other fields. T-LiDAR uses a coherent laser beam with very little divergence caused by stimulated emission. It is a non-contact, non-destructive and non-penetrative active recording system which is stationary during the recording. The electromagnetic waves are reflected by surfaces and the receiver detects portions of the backscattered signal, hence, the two-way travel time is recorded. Later, point measurements from different spatial directions are used to compute the object geometry. After implementation of the dataset in a geographical information system (GIS), accurate digital elevation models (DEM) or digital terrain models (DTM) can be generated. In this study the laser ranging system ILRIS 3D from OPTECH Inc., Ontario, Canada (fig. 2.9) was applied equipped with a digital camera that allows combining the point cloud with panchromatic information. The advantages of the terrestrial method are the flexible handling, a relatively quick availability of an actual dataset, and a very high spatial resolution of the object with information about intensity, x-y-z-coordinates and range. The quality of the reflection depends on the inclination angle of the laser beam, the range between the object and scanner, the material, the colour, the surface condition (weathering/roughness), and the spatial resolution. For this study the distance between the LiDAR and the fault scarp ranges between 5 and 10 m, thus offering a maximum resolution of 1 mm.

Figure 2.9.: ILRIS 3D system from OPTECH Inc. Photograph by Thomas Wiatr.

2.5.2. Geomorphological Analysis

More than 30 profiles of fault scarps were generated from scarps found in limestones and harzburgites and a few found in sandstone and serpentinite. Field measurements were carried out by laying 1 m wooden poles on the slope surface perpendicular to the strike of the fault plane. The wooden poles were arranged along a tape measure which was adjusted to have the same dip direction for all measurements. The inclination of the profile was then measured on each wooden pole at 1 m intervals with a compass clinometer. In order to reconstruct the offset, a minimum of 40 m length was considered. If less than 40 m is used, the graphical construction of the offset in the profiles is not possible as seen from data taken from very first profiles and according to Papanikolaou et al. (2005). Therefore, it was necessary to exceed a length of 20 m above and below the scarp face. GPS data were taken at the start and end point of the profile, as well as at the top edge of the scarp, in order to tie in the profile. Only natural scarps were measured that have not been anthropogenically influenced. At locations where an anthropogenic influence could not be avoided, the profile inclination was extrapolated. For the final analysis only scarps were taken into consideration, that are not subject to unusually high erosion or sedimentation rates (e.g. sandstone or serpentinite scarps). Profiles were plotted and graphically evaluated.

In general a scarp can be divided into a number of sections (see fig. 2.10: upper slope, degraded scarp, free face, colluvium and lower slope). These sections were defined in each profile and the mean dip of the upper and lower slopes was determined. To construct the thickness of the colluvium and the theoretical edge of the scarp before erosion, the slopes were geometrically extended towards the free face. From this it was possible to calculate the theoretical offset along the rupture. The thickness of the colluvium was observed and estimated in the field. Using the data from these profiles, a complete geometry of diverse fault scarps measured all over the basin can be obtained.

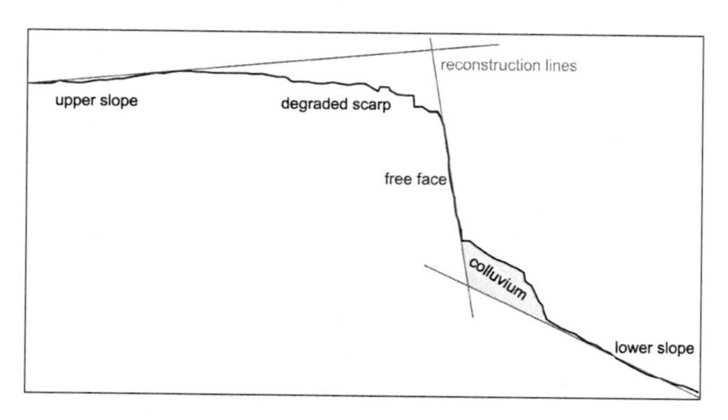

Figure 2.10.: Scarp profile with denomination of sections.

Possible sources of error for the construction of scarp profiles are:

- Although the dip direction of the wooden poles was checked every meter, small variations in dip direction may be possible. The effect on the inclination data will be negligible over the complete profile as only very minor variations could have been measured.

- The profiles were measured perpendicular to the free face. However, as the scarp is changing its direction over its total length, the general direction of strike of the fault trace was taken as a measurement. Therefore, some of the profiles differ slightly from the strike of fault planes plotted on stereonets.

In addition to profile construction remote sensing data were used to trace faults, evaluate fault lengths and segmentation, to identify and to quantify other seismogenic features. To quantify the stages of dissection and to determine the activity along the triangular facets found in the southwest and northeast of the basin, mountain front sinuosity indices were calculated by using the method proposed by Bull (2007). The sinuosity of the mountain-front is defined by the the ratio of the length of the mountain-piedmont junction L_j to the length of the range-bounding fault L_s (fig. 2.11.

$$J = L_j/L_s$$

This provides information about the evolution of the geomorphic system to a base level fall induced by active faulting. Other geomorphic indices were not taken into account as the different lithologies produce such high uncertainties, that the definition of activity classes throughout the basin did not produce reliable results.

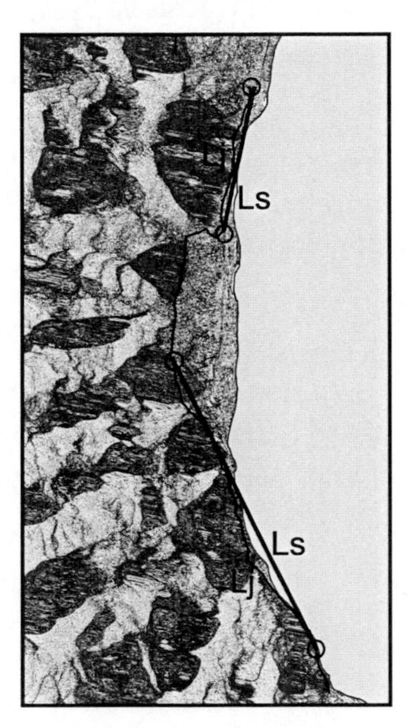

Figure 2.11.: Tectonically active mountain front associated with an oblique right-lateral fault in the south west of the Lake Ohrid Basin. The length of the thick, straight line L_s, is the length of the range-bounding fault. The thin, sinuous line, L_j, is along the mountain-piedmont junction. Its sinuosity records embayments at mouths of watersheds, and minor departures from fault-zone linearity.

2.5.3. Workflow

The work at the Lake Ohrid Basin showed that if using a variety of different methods it is necessary to adhere to a workflow. The workflow proposed here is a good foundation for neotectonic studies in the Mediterranean or related climatic conditions and with similar objectives as in the Ohrid Basin.

To gain an overview of the working area and to define target areas the use of satellite images, DEMs, and all kinds of geological maps is crucial. In this first approach an initial idea of the working conditions is developed and interesting areas for further studies can be defined. Besides that geomorphic features can be identified and quantified and accessibility issues must be sorted out. In the next stage structural and palaeostress analyses are suitable to define the strain of the region, to get an idea of the successive deformational events in the working area and to gain information of the recent stresses and deformational mechanisms.

Geophysical studies must be carried out to investigate larger areas in order to obtain information on sedimentary structures and deposition environments. In the Lake Ohrid case, ERM profiles in Wenner array delivered the best results. GPR measurements with the 270 MHz delivered good results especially in combination with percussion drilling, whereas the 100 MHz antennas resolution is not high enough to determine fine sedimentary structures. Besides that the high water table and saturated soils let the signal diminish in the upper layers, so that no data was obtained below 10 m. Accompanied by shallow drilling operations and further analysis of sediment structures, the geophysical measurements delivered satisfying results.

Accompanied by the results of the palaeostress studies geomorphological analyses are a powerful tool to tie in the deformational history. Profiling of identified scarps delivers geometries of faulting and throw rates. The collected field data are then processed and completed by a detailed geomorphological analysis using remote sensing techniques and calculating slip rates.

In the successive field trips a trenching study would have provided good data. The spotting for suitable trenching locations was not successful due to rough ground conditions and the inaccessibility of sites because of housing or other infrastructure. However, trenching would accomplish our studies and give a better understanding of recurrence and slip rates (cumulative and per event).

3. Regional Setting

The Lake Ohrid Basin is located on the Western Balkan Peninsula and belongs politically to the Republic of Albania (Republika e Shqipërisë) and the Republic of Macedonia (Република Македонија). The basin is bounded to the east by the Galicica Mountain Range with heights of up to 2,250 m, and to the west by the Mokra Mountain Range with heights of up to 2,200 m; these confine the research area around Lake Ohrid (see fig. 3.11). These mountain ranges are part of the collision zone between the Eastern Alps and Western Turkey forming the Alpine-Carpathian-Dinaridic orogenic system (see fig. 3.1 for tectonic units).

3.1. Geodynamic setting

The geodynamics of the Western Balkan Peninsula are today mainly controlled by the systems of the Northern Hellenic Trench and the North Anatolian Fault Zone (NAFZ; fig. 3.2; Armijo et al., 1999; McKenzie, 1972; Papanikolaou et al., 2006). Within this area, the Lake Ohrid Basin forms one of the largest of a number of basins in the NW-trending Dinaride-Albanide-Hellenide (see fig. 3.2) orogen that stretches parallel to the eastern Adriatic coast. The definition of the orogen and its subdivisions is not clear and varies in the literature (see fig. 3.3 for an overview) depending on the focus of the research and the background of the authors. Here, the subdivisions defined by Kotzev et al. (2008) are used.

The eastern margin of the Adriatic Sea has undergone dramatic changes since the Palaeozoic. It is part of the very complex collisional tectonic system of the Mediterranean. It involves fragments of all sizes and shapes, originating from Western Gondwana and traveling northward. Ocean basins evolved, controlled by the interaction of individual blocks with each other and with the super continents of Laurasia and Gondwana. After research carried out by Dilek (2006), the evolution of the Tethyan systems (Palaeo- and Neo-Tethys) turned out to be much more complex than considered in previous geodynamic models. Here, the main phases concerning the Dinaride-Albanide-Hellenide orogen and the national territory of Macedonia are described.

The Adriatic-Dinaridic carbonate platform was subject to an extensive rifting phase initiating in the Late Permian resulting in the subsidence of the rift shoulders and establishment of a shelf. In the Middle Triassic the crustal separation came to a halt. In the Late Triassic/Early Jurassic, the opening of the Dinaridic branch of the Tethys started, probably under pelagic conditions. This period of sea-floor spreading lasted until the Late Jurassic (Dilek, 2006; Pamić et al., 1998; Schmid et al., 2008; Tari, 2002). Then (Late

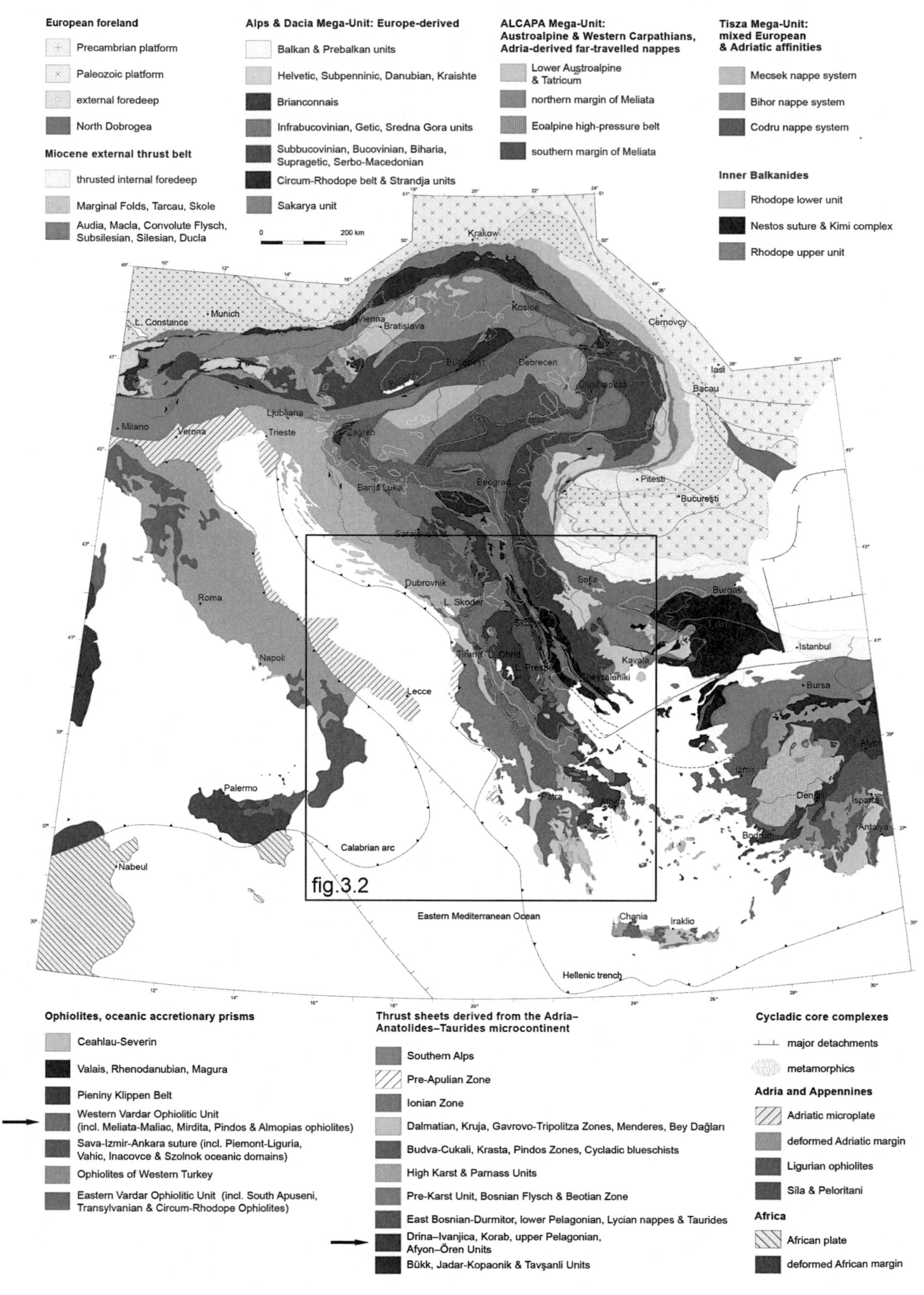

Figure 3.1.: Tectonic units of the alpine collision zone between the Eastern Alps and Western Turkey. Modified after Schmid et al. (2012).

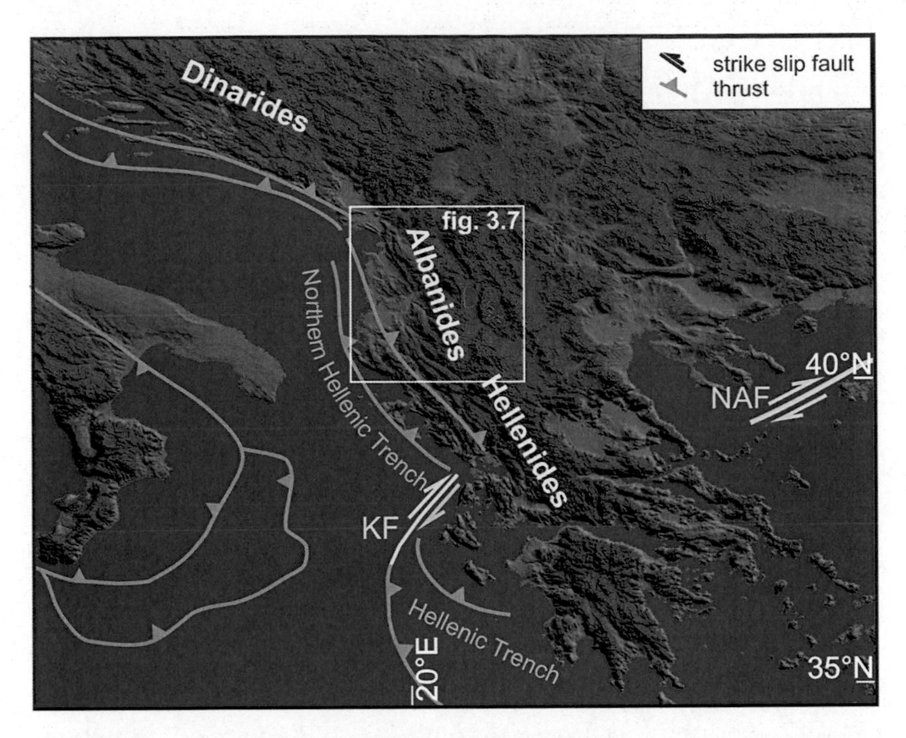

Figure 3.2.: Overview of the eastern Mediterranean geodynamic situation. Inset shows study area in figure 3.7. KF = Kefalonia Transform Fault, NAF = North Anatolian Fault. Modified after Hoffmann et al. (2010).

Jurassic/Early Cretaceous) an initial subduction was established with modern Macedonia located on the hanging wall of the subduction zone leading to the gradual closure of the Vardar Ocean. A magmatic arc was then established along the active continental margin. The ongoing subduction is expressed in widespread thrusting, folding and strike-slip faulting (Dumurdzanov et al., 2005). By the end of the Late Eocene, the strongest tectonic movements mark the collision of the Serbo-Macedonian and Rhodope-Pelagonian blocks (Dilek, 2006; Pamić et al., 1998). In this main deformation phase the NW-SE trending folds and thrusts formed in the north (Dinarides), most likely by north or northeast dipping subduction. To the south (Albanides/Hellenides) E-W trending transtensional basins opened (Dilek, 2006; Pamić et al., 1998). In the Eocene NW-trending basins evolved in Eastern Macedonia in an extensional regime probably related to rollback at the trench and/or gravitationally induced spreading within the magmatic arc (Dumurdzanov et al., 2005). In Central Macedonia a forearc basin developed. This was followed by NW-trending folds to the west formed by Late Oligocene to Early Miocene shortening. This period could have been caused by the subduction of the Kruja fragment which may have led to the slowdown of the subduction rate and a stress regime change from extension to compression in the overlying crust (Dumurdzanov et al., 2005).

The roll back of the subducted slab (fig. 3.4) over time led to a westward migration of the entire geodynamic system. Slab roll back is generally associated with uplift as evidenced in the Gibraltar Zone (Duggen et al., 2003) and shown by the fission-track dates of Muceku et al. (2008). In this case it is evidenced by the westward migration of the E-W extensional domain of Eastern Macedonia which was established in the Early Miocene and has been the prevailing deformation mechanism until present (Dumurdzanov et al., 2005). The extensional deformation is probably related to gravitational lateral spreading

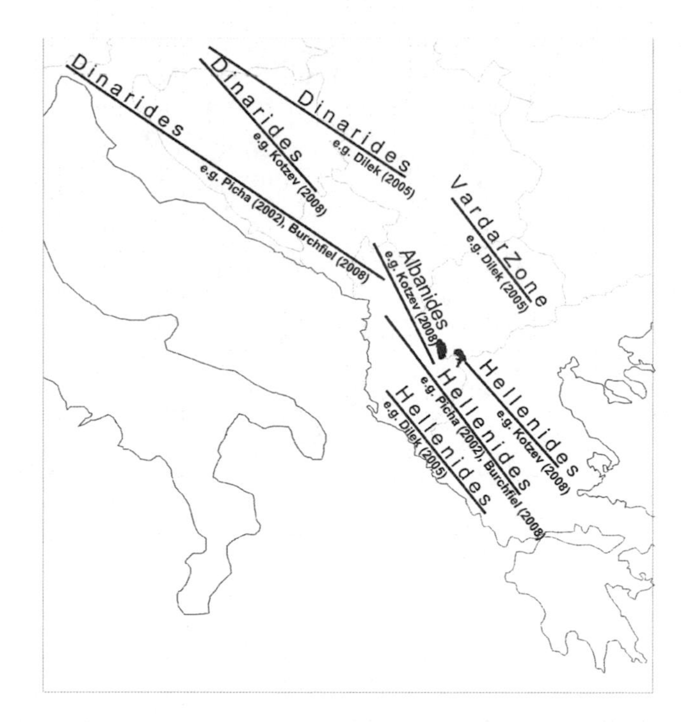

Figure 3.3.: Definitions of the Dinaride-Albanide-Hellenide orogen by different authors.

of thickened crust together with trench rollback. The formation of the Ohrid Basin began in the Late Miocene as a pull apart basin controlled by right-lateral strike-slip movements. Subsidence and further extension account for the major dynamic component which has been occurring since the Pliocene-Pleistocene (Dumurdzanov et al., 2005).

In the Early Pliocene the NAFZ (fig. 3.2) started to extend into the Aegean and to establish right-lateral slip (Burchfiel et al., 2008; Dumurdzanov et al., 2005) causing a major tectonic change. This can be seen in Eastern Macedonia where older N-trending basins are now disrupted by E-W trending basins. Faults become younger to the west due to westward migration of the trench rollback (Dumurdzanov et al., 2005).

Presently, no oceanic crust is subducted below the Adriatic foreland north of the Kefalonia transform fault or at least subduction rates are very slow (Yanev et al., 2008). The collision zone between the northward moving Afro-Arabian plate and the Eurasian plate shows convergence rates between 10 mm/a (Cyprus Arc) and more than 40 mm/a (Hellenic Arc; Dilek, 2006). Today, the main structural sections of the Eastern Adriatic coast can be subdivided into a compressional coastal domain, followed by a narrow zone of transition west of Lake Ohrid; the extensional domain is located further east in which the Neogene basins formed (see fig. 3.4; Dumurdzanov et al., 2005).

Within this system the Ohrid Basin, the Debar Basin to the north, the Korca and Erseka Basins to the south, and the lakes of Prespa and Mikri Prespa to the southwest are situated in a basin and range-like geodynamical setting (figs. 3.4 and 3.7). While the Ohrid Basin is a graben structure, the associated Korca and Erseka Basins are halfgrabens bordered by a NW-SE trending normal fault on their eastern side. The entire area is controlled by present day E-W extension (fig. 3.4). In this structural setting, the Lake Ohrid Basin is situated on top of the lithological boundary between the Palaeozoic orogen in the east (Pelagonian) and the Mesozoic rocks (Apulian) in the west (Robertson, 2004), or per definition after (Schmid et al., 2008, see also fig. 3.1), between the Western Vardar Ophiolithic Unit (Mirdita) and the Korabi Unit . Jozja and Neziraj (1998) and Tremblay et al. (2009) defined these units as the Mirdita Ophiolite and the Western Macedonian Zones (see also chapter 3.3).

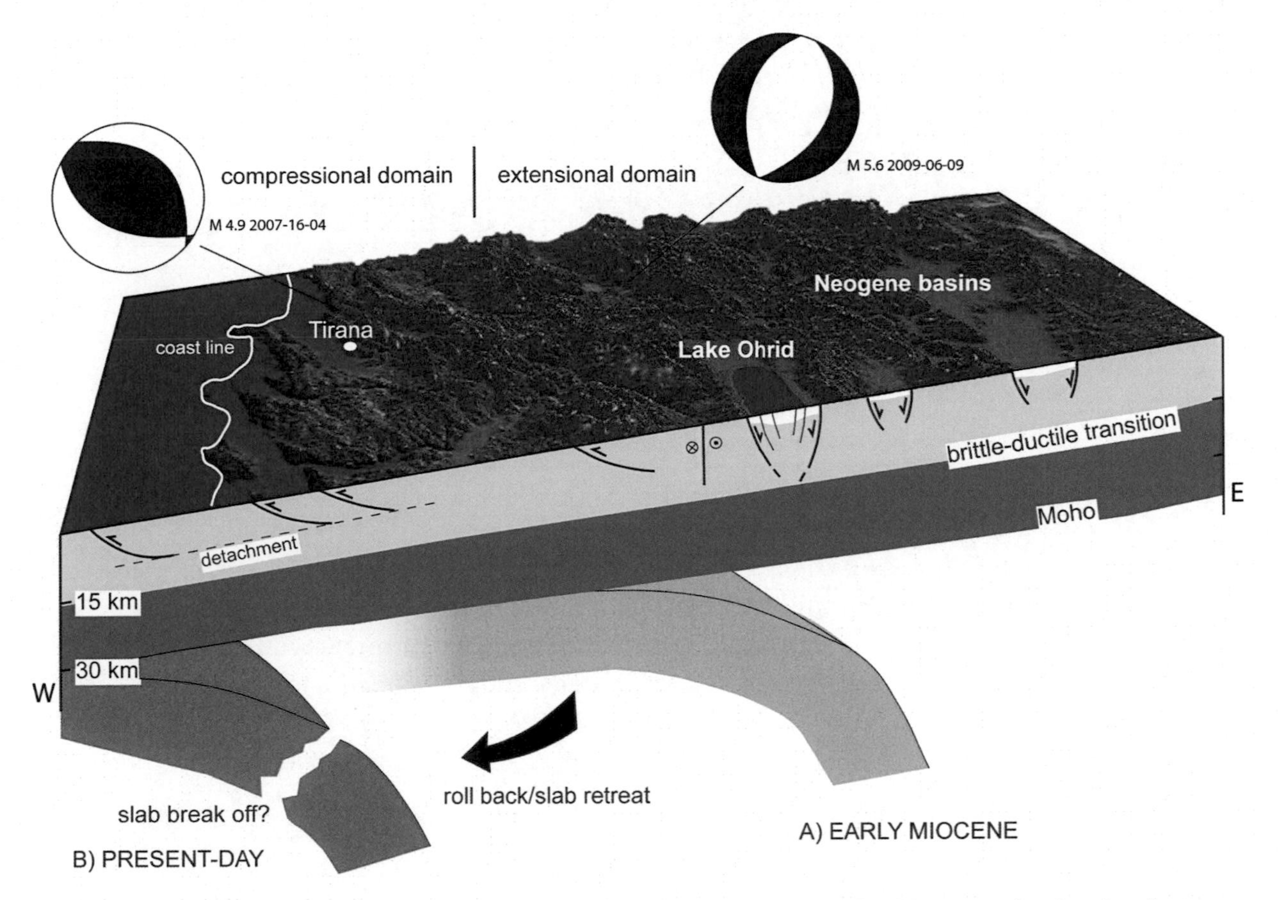

Figure 3.4.: Structural cross section from the Adrian coast to the Neogene basins in the Balkanides. The frontal part is characterised by thrusts, whereas the basins are formed within the extensional domain. Note subduction roll-back since Late Miocene. The Moho dips eastward from 30 km to about 40 km depth (Anderson and Jackson, 1987; Grad et al., 2009; Milivojevic, 1993). Modified from Hoffmann et al. (2010).

Currently, the Lake Ohrid Basin is flanked by active N-S trending normal faults (with variations of c. 20° NW and NE). Fault surfaces and lineations are preserved in the present-day landscape as hard rock fault scarps (mainly in limestones and harzburgites). These morphological features tend to trend mainly N-S in the west of the basin and N-S to NNE-SSW in the east. Other sets of NW-SE and E-W lineaments are also present (Wagner et al., 2008). The deformational forces behind the morphological expressions can be derived from recent earthquake data (see chapter 3.2) and by palaeostress analysis (see chapter 4).

GPS velocity models from Burchfiel et al. (2006) and Caporali et al. (2009) show a uniform SSE migration of Macedonia together with western Bulgaria as a joint crustal piece with 3-4mm/a (see fig. 3.5). Burchfiel et al. (2006) estimate present day slip-rates of max. 2mm/a on "NNW-striking normal faults and associated strike-slip faults with right-lateral displacement" but with a very high uncertainty due to a coarse-meshed GPS network. At

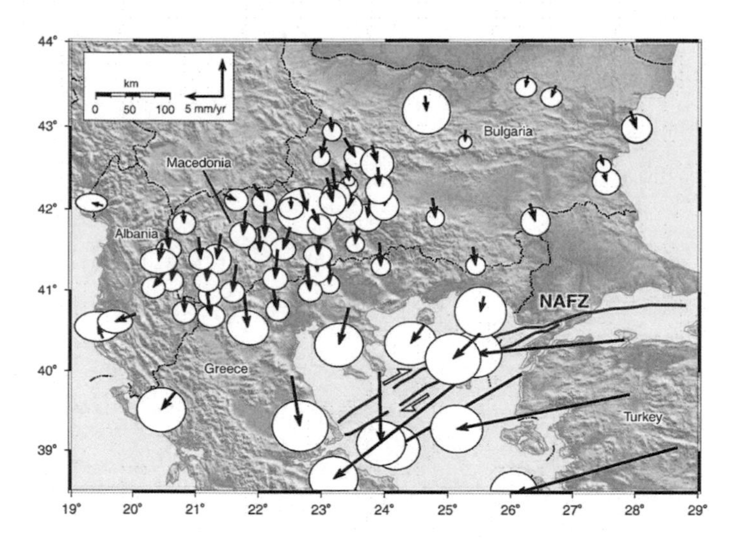

Figure 3.5.: GPS velocities with respect to the Eurasian reference frame defined by McClusky et al. (2000). Uncertainties are shown at 95% confidence. From Burchfiel et al. (2006).

the same time Adria moves N-NE with respect to Eurasia. Figure 3.6 shows a velocity profile across Albania, Macedonia and Greece crossing also Ohrid and Prespa lakes. The right lateral shear zone is the consequence of the oblique convergence along the Dinaride-Albanide-Hellenide orogen resulting in strain partitioning. The velocity drop of 8.7 ± 0.5 mm/a eastwards is a measure of the strain rate associated with the subduction of the Apulia platform under the Albanides (Caporali et al., 2009; Dilek, 2006).

3.2. Seismicity and Neotectonics

The Lake Ohrid Basin is an important N-S trending graben structure and is located within the Afro-European Convergence Zone, a region with moderate seismicity (fig. 3.7). The central mountain chain, especially the intramontane basins of Late Neogene age and the plate boundary along the coast of the Adriatic Sea, form one of the active seismic zones in Albania/Macedonia with frequent moderate earthquakes having occurred in the last few centuries (Aliaj et al., 2004; Goldsworthy et al., 2002; Muco, 1998; NEIC, 2013). In contrast to the compressive western part of Albania, the central and eastern parts of Albania as well as the west of Macedonia are presently subject to extension as evidenced by fault plane solutions of earthquakes (fig. 3.7; Burchfiel et al., 2006). Earthquake focal mechanisms show active N-S normal faulting with horst and graben structures in a basin and range like environment.

Major earthquakes occurred during historical times. In AD518 an earthquake destroyed the cities of Ohrid and Skopje (110km NNE of Ohrid) (Petrovski, 2004). Ambraseys (2009) states that this earthquake was located in the province of Dardania (for location see figure 3.8). The event was so strong that almost the entire city of Ohrid had to be rebuilt; Skopje also suffered heavy damages. The contemporary chronicler Marcellus Comes writes that "24 castles or villages were ruined by repeated earthquake shocks. [...] The metropolitan city of Scupi (near Skopje) was ruined to its foundations. [...]

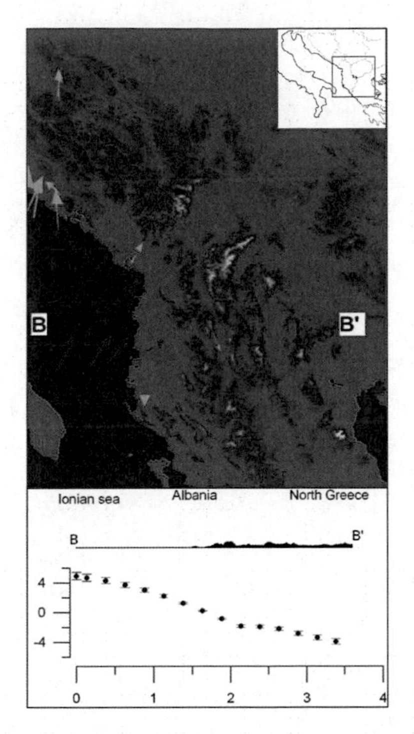

Figure 3.6.: Velocity profile across Albania, Macedonia and Greece. The profile BB' shows a large scale right lateral shear zone in the Balkans area to be accommodated by right lateral strike-slip faults in Albania and Western Greece. The cross track velocity is plotted. Units for the x-axis are 100 km, for the y-axis mm/a. Modified from Caporali et al. (2009).

Mountains [...] were rent asunder, rocks and forest trees were torn from their sockets; and openings in the ground twelve feet in breadth and 30,000 Roman feet [43 km] intercepted." (Ambraseys, 2009). This description suggests an event of a minimum intensity of X or XI and a magnitude of 7 or more (according to Wells and Coppersmith, 1994). This is believed to be the strongest historical earthquake to have hit Macedonia.

Another strong earthquake in the study area is reported by the ancient historian Procopius of Caesarea (ca. AD500-565) who mentioned Lychnidos (in Greek Λυχνιδøς, in Latin Lychnidus) in his Secret History or History Arcana (Atwater, 1927). Lychnidos is regarded as the old name of Ohrid, but there are also some authors that place it close to Sveti Naum at the southern shore of the lake (Lempriere, 1838; Malte-Brun, 1929). It was completely destroyed by an earthquake on May 29th or 30th in AD526. Atwater (1927) states:

"Who could number those that perished in these metropoles? Yet one must add also those who lived in [...] Lychnidus in Epirus; and in Corinth: all thickly inhabited cities from of old. All of these were destroyed by earthquakes during this time, with a loss of almost all their inhabitants. And then came the plague which I have previously mentioned, killing half at least of those who had survived the earthquakes. To so many men came their doom, when Justinian first came to direct the Roman state and later possessed the throne of autocracy". The city was probably rebuilt by Emperor Justinian (AD527-565) who was born in the vicinity, and was named Justiniana Prima by him; this was the most important of the several new cities that bore his name.

The text of the Secret History Arcana is an emotional harangue against emperor Justinian

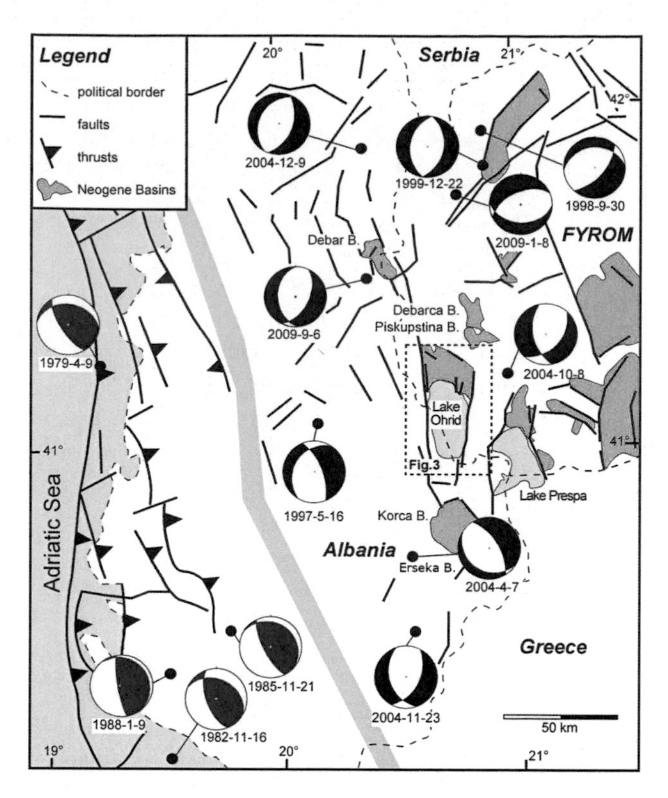

Figure 3.7.: Fault plane solutions of several earthquakes in the triangle of Albania, Macedonia and Greece. Black beachballs indicate normal faulting, red beachballs indicate thrust faulting (source NEIC (2013); CMT focal mechanisms). Grey zone divides compressional and extensional domains associated with Neogene basins and normal faults (modified from Dumurdzanov et al., 2005). See section in figure 3.4 for structural trends. Modified from Reicherter et al. (2011).

and his wife, and describes the catastrophe during Justinian times. This earthquake is linked to the Antioch earthquake on May 29th AD 526. A local earthquake that destroyed the entire city and left the majority of the inhabitants dead must have had a magnitude greater than 6, even taking into account poor building standards and historical (or political and personal) exaggerations. Even if the fatalities were not caused by building collapse due to shaking but resulted from secondary seismic effects like rockfalls, landslides, wells running dry, etc., a medium-strong event (Michetti et al., 2007) must be assumed.

Ambraseys (2009) lists a strong event in AD 527 which would fit the historical data during the reign of Justinian. Ambraseys (2009) links the AD 527 earthquake with another event, reported by various authors between AD 528 and AD 539, at a site north of Niš in Serbia. It is not clear whether these events are related to the earthquake of AD 518 which destroyed Ohrid and Skopje, or if, due to historical uncertainties, only one event took place. Other significant events occurred in February AD 548; April 3rd 1673, February 7th 1871, September 10th 1889, and September 28th 1896 which was M 6.7 according to Papazachos and Papazachou (1997); (Ambraseys, 2009; Ambraseys and Jackson, 1990; Goldsworthy et al., 2002).

Instrumental seismicity records in the Ohrid area reach back to the early 20th Century. The strongest event ever measured here took place on February 18th 1911 at 9.35 p.m. This magnitude 6.7 earthquake (corresponding to EMS X) occurred in the Ohrid-Korca

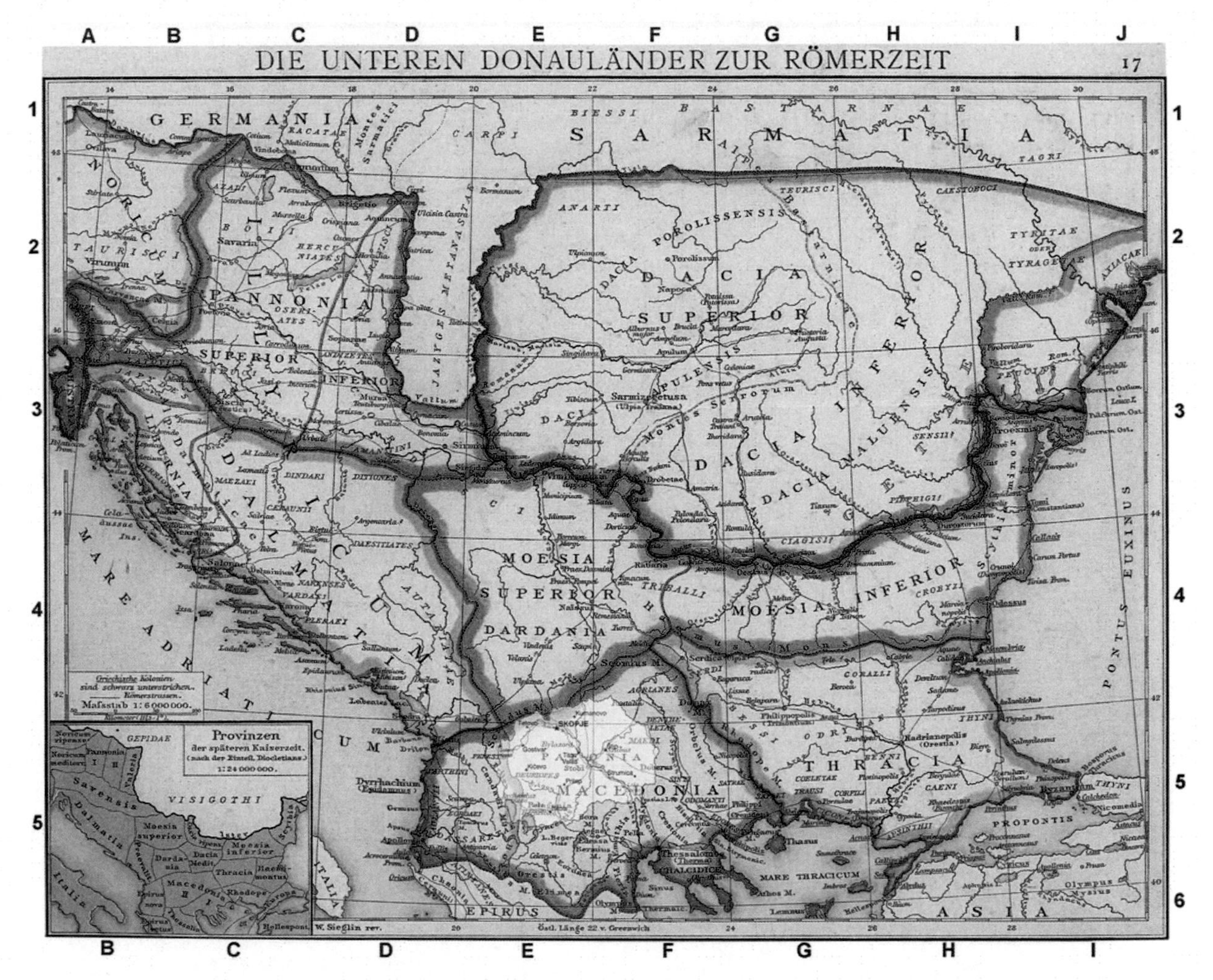

Figure 3.8.: Location of Dardania (marked in yellow) during Roman Age. Grey shaded region shows the area of modern Macedonia. From Andree (1886).

area (40.9°N, 20.8°E) at a depth of 15km (Milutinovic et al., 1995; Muco, 1998).

The most recent events are the November 23rd 2004, M_w 5.4 earthquake in the Korca region (40.39° N, 20.48° E; focal depth ~20km, normal faulting), the September 6th 2009, M_w 5.6 event in Debar, Albania (41.46° N, 20.41° E; focal depth ~2km, normal faulting) and the June 8th 2012, M_b 4.4 earthquake 15 km northeast of Ohrid city (41.24° N, 20.90° E; focal depth ~10km, normal faulting; EMSC, 2013; NEIC, 2013). During the September 6th 2009 earthquake, hundreds of houses were damaged and some dozens were destroyed; however, no fatalities were reported. A series of more than 35 aftershocks followed the main shock. The shallow epicentres in combination with the poor building standards in the region are held responsible for the severe damage. Events with magnitudes below M_w 5 occur more frequently (e.g. February 8th 2012, M_w 4.0, 40.81° N 20.53° E; November 26th 2012 M_L 2.9, 40.88° N 21.11° E; EMSC, 2013).

Burton et al. (2004) list only moderate events (except the 1911 earthquake) at shallow depths (< 60km) within the study area. Background seismicity is low compared to Greece, Western Albania and the Eastern Macedonia-Bulgaria region. Hypocentre depths scatter between 10 and 25 km, but some deeper earthquakes occur between 25 and 50 km depth. Very rarely intermediate earthquakes around 100 km depth are observed. Small and mod-

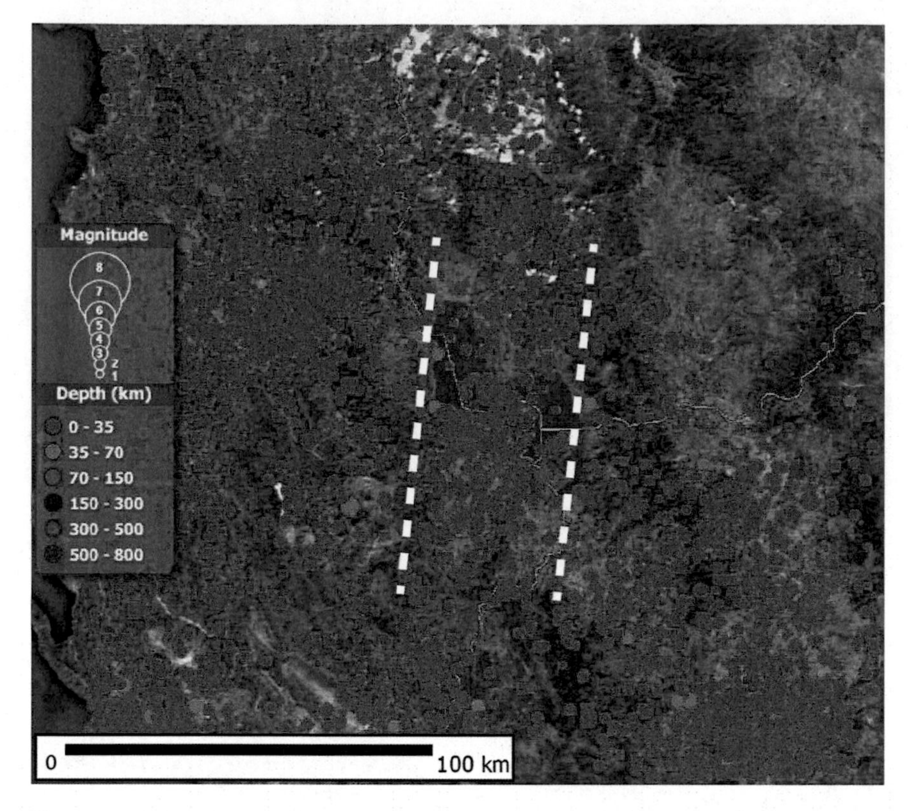

Figure 3.9.: Seismicity in the triangle of Macedonia, Albania and Greece. Data are derived from NEIC (2013) and draped on Google Earth. Dashed lines show the corridor of seismicity in the Ohrid-Korca zone.

erate earthquakes (e.g. the <M_b 4.4 Ohrid earthquake 2012) take place predominantly along major fault zones, and are concentrated along the margins of the Lake Ohrid Basin. The Ohrid-Korca Zone is considered to be the region of the highest seismic hazard in the Albanian-Macedonian corridor based on present-day seismicity (see fig. 3.9; Aliaj et al., 2004).

Maximum peak ground accelerations expected for a return period of 500 years reach from 0.30g in the east of Ohrid (Lake Prespa region) to 0.50g in the west of the study area (Iben Brahim, 2005). Besides segmentation of longer faults, and taking into account a possible rupture length of approximately 10 - 20km, it is expected that the normal faults at Ohrid are prone to generate an earthquake of M 6.5 - 7.0 (these estimations were made according to the theories of Wells and Coppersmith, 1994). However, recurrence periods of major events in the Lake Ohrid area are unknown. Taking all this into account, the Ohrid-Korca area is regarded one of the highest seismic hazard zones in Albania and Macedonia (Aliaj et al., 2004) but only medium among the Balkans (GSHAP, 2013; NEIC, 2013).

3.3. Regional Geology

The Lake Ohrid Basin graben structure is located at the contact between the Korabi Zone (part of the Western Macedonian Zone after Dumurdzanov et al. (2004), or the Pelagonian

Figure 3.10.: Lithological map of the Lake Ohrid area. Compiled after Dumurdzanov and Ivanovski (1977); Premti and Dobi (1994).

units after Robertson (2004) and the Mirdita Ophiolite Zone (Adriatic/Apulian units after Robertson (2004); Shebenikum Massif after Jozja and Neziraj (1998).

In the lake region the Jurassic ophiolites of the Mirdita Zone underlie the units of the Korabi Zone (fig. 3.10), which were thrust in a westward direction during Early to Mid Tertiary (Robertson and Shallo, 2000). The contact between the two zones can be observed south of the Lini Peninsula (figs. 3.11) at the western shore of the lake. The geodynamic evolution in the Dinaride-Albanide-Hellenide thrusting system led to a NW-SE trend of the large geological and structural units.

Lithologies around Lake Ohrid mainly comprise low to medium-grade metamorphosed Palaeozoic sedimentary rocks at the NE margin of the basin and Triassic carbonates of the Galicica and Mali i Thate Mountains (Dumurdzanov et al., 2005; Hoffmann et al., 2010). The SW graben shoulder is characterised by Jurassic ophiolites of the Mirdita Zone, mainly consisting of lherzolites, gabbros, minor harzburgites and Cretaceous limestones. Syn- and post-orogenic Oligocene-Miocene molasse sediments are only present to the west and south of Lake Ohrid (see fig. 3.10; Hoffmann et al., 2010).

3.3.1. Korabi Zone

The Korabi Zone (fig. 3.10) in the lake area is characterised by Palaeozoic, mostly metamorphic and magmatic rocks which are superposed by Triassic to Early Jurassic limestones (Kilias et al., 2001; Robertson and Shallo, 2000) in the horst shape of an anticline structure between Lake Ohrid and Prespa (fig. 3.10). The Devonian metamorphic rocks consist mainly of greywackes (metasandstone) and phyllites (sericite shales) with a complex metamorphic history. These metamorphic rocks crop out at the northeastern lake shore around the city of Ohrid. South of the city of Ohrid they are preserved as a tectonic window due to extensive normal faulting (fig. 3.10). In the northern part of the Galicica Mountain range, marbleised limestones are preserved locally at the top of the mountains, e.g. northeast of the village of Kosel (fig. 3.11). Volcanic activity in this unit is only preserved as Palaeozoic granitic intrusions north of Lake Prespa (Dumurdzanov and Ivanovski, 1977). These units are superposed by Triassic to Early Jurassic platform carbonates of the Galicica Mountains which consist mainly of limestones and locally of dolomites. The mainly massive and thick limestones are locally thinly-bedded and intercalated with radiolarites and cherts, e.g. northeast of Ljubanista (fig. 3.11). These carbonates of the Galicica Mountains extend to the south of the Mali i Thate Mountain chain (fig. 3.11) and are affected by intense karstification. Most parts of the eastern graben shoulder and of the northwestern shores of the lake are built up by these Mesozoic units. Along prominent normal faults, like the one at the village of Dolno Konjsko (fig. 3.11) south of the city of Ohrid, serpentinites are exposed as isolated blocks in shear lenses. Their origin and stratigraphic position has not been clearly understood. Furthermore, Mesozoic intrusions of rhyolithes and diabases are preserved in between the limestones and dolomites east of Kosel (fig. 3.11).

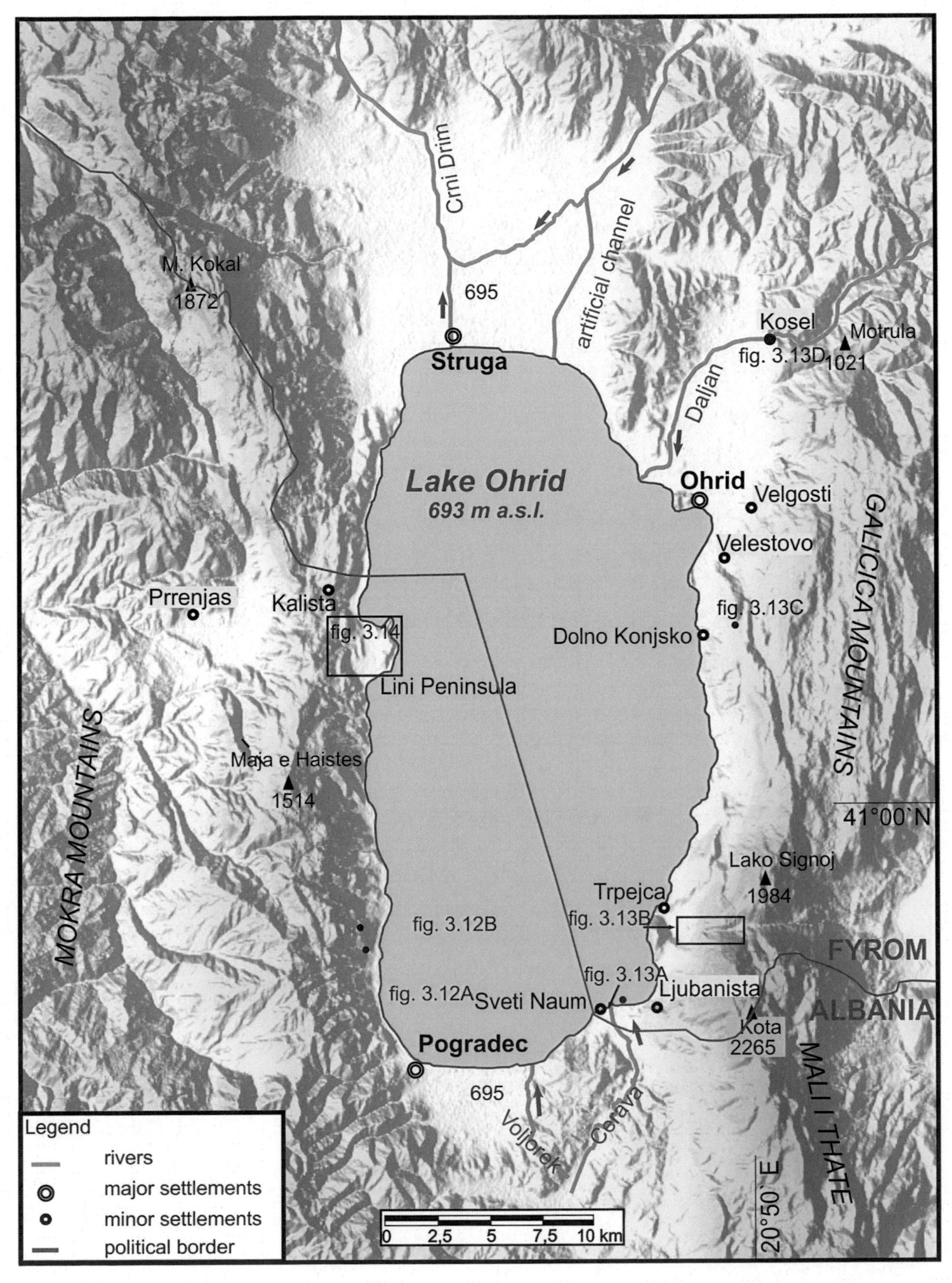

Figure 3.11.: Overview of the Lake Ohrid research area. Red marked areas indicate the locations of figures 3.13, 3.14 and 3.12. Modified after Hoffmann et al. (2010).

3.3.2. Mirdita Zone

The Shebeniku Ophiolite Complex of the Mirdita Zone (fig. 3.10) belongs to the eastern ophiolite belt (Kocks et al., 2007). The genesis and origin of the sub-parallel western and eastern ophiolite belt of the Mirdita Zone is still under discussion (Gawlick et al., 2008; Kilias et al., 2010; Muceku et al., 2006; Robertson, 2004). While west-directed transport of the nappe structures in the Tertiary is generally accepted, their origin as far travelled allochthonous nappes (Gawlick et al., 2008; Kilias et al., 2010) with origin in the Vardar Zone, or autochthonous development in an independent Pindos-Mirdita Ocean (Robertson and Shallo, 2000) are still under debate. The ophiolites of the eastern belt are of Early to Middle Jurassic age (fig. 3.10) and show supra-subduction zone (SSZ) and mid-ocean ridge (MOR) affinities in accordance with the models of Bébien et al. (1998) and Hoeck et al. (2002). They consist mainly of lherzolites and gabbros, interlayered pyroxenitic dykes, harzburgites and serpentinites (Hoeck et al., 2002; Kocks et al., 2007).

Figure 3.12.: Panoramic view of the Lini halfgraben (Albania; view towards southeast), note the stepped landscape due to normal faulting of the ophiolites **O** and Late Cretaceous limestones and Triassic carbonates **C**. From Hoffmann et al. (2010).

These ultrabasic rocks are covered by lateritic horizons bearing Fe-, Ni- and Al- ore deposits. Above the sequence basaltic breccias and turbidites with conglomerates of ophiolitic clastics are deposited followed by Cretaceous limestones. To the west, the ultrabasic units of the Shebeniku Ophiolitic Massif are transgressively covered by thin shallow water carbonates of Late Cretaceous age (figs. 3.10 and 3.12) containing rudists (Jozja and Neziraj, 1998). This transgression is characterised by an angular unconformity. Figure 3.13A shows the contact of deeply weathered lateritic horizon and the shallow water limestones. As the ore deposits in the lake area were intensely mined in the past this contact is often indicated by linearly orientated tailings from mining operations. In any case, these contacts are not always sedimentary and uniform. Due to intensive normal faulting during the extensional stage, the deposits are arranged in staircase-like expressions and today the contacts are often disturbed (figs. 3.12 and 3.13B).

3.3.3. Syn- and postorogenic development

During alpine orogeny from the Eocene until Pliocene, flysch and molasse-like sediments were deposited which are now exposed as deformed nappes (Jozja and Neziraj, 1998) and cover the ophiolitic Mirdita units along the western shoreline. Further they are preserved covering discordant Korabi units to the south of the lake (fig. 3.14A). They consist of Pliocene flysch comprising conglomerates, silt- and sandstones overlain by Pleistocene

Figure 3.13.: A: The contact between Mirdita ophiolites and Late Cretaceous limestones is marked by large tailings (west coast of Lake Ohrid, Albania, view towards SW). B: Close-up view of a normal fault that bounds the lateritic horizons of the Mirdita ophiolites and the limestones of Late Cretaceous age (Albania, 40° 57′ 48″N, 20° 36′ 22″E; view towards N). From Hoffmann et al. (2010).

molasse clastics, comprising mainly conglomerates, sands and boulder gravels. Molasse outcrops at the lake are located near Pogradec (Miocene-Pliocene) and west of Prrenjas (fig. 3.11; Eocene to Tortonian), and consist of folded and thrusted granite-bearing conglomerates and sandstones. A further outcrop is located near Ljubanista, close to the southern shore of Lake Ohrid (fig. 3.14A) where the river Cerava cuts into undeformed Pliocene conglomerates consisting of sands and gravels. The deposits are built up of eroded Cenozoic molasse of the Korca Basin which have been redeposited in the Ohrid Basin. They are superposed discordantly by Pleistocene conglomerates which have been transported from the heights of the Mali I Thate and Galicica Mountains. The transportation processes were stopped by ongoing subsidence which divided the Ohrid from the Korca Basin in the Late Pliocene and Early Pleistocene.

To the SE of Trpejca, Pleistocene carbonate-cemented coarse-grained angular colluvial sediments form a large and well-preserved debris cone (fig. 3.14B). Other areas are characterised by the formation of huge palaeosols which are, for example, preserved within the hanging wall of an active normal fault at the eastern graben shoulder NW of Dolno Konjsko (fig. 3.14C; see chapter 6.2.1 for details). The youngest deposits in the Ohrid Basin are the Quaternary plains of Struga in the north and Pogradec in the south. While the plain of Pogradec provides indications for a drying up of shallow lake areas filled with fine grained sediments, the northern plain is built up by gravel and sand strata from river deltas and alluvial fans. North of the city of Ohrid in the village of Kosel the so called Duvalo "volcano" can be observed. The fault-related hydrothermal field with carbon dioxide and hydrogen sulphide exhaling solfatara or fumaroles (fig. 3.14D; geochemical data are not available) is situated some kilometers north of the town of Ohrid and stretches along a N20E striking lineament (Arsovsky and Hadžievsky, 1970) in highly alterated phyllites of Devonian age which are kaolinisated. These rocks bear sulphur which was mined and used for spa and therapeutic purposes. Today, the hydrothermal field is used as a dumpsite and thus the solfatara are successively buried. Further south along the lineament, thermal sulphur-bearing springs occur in the village of Velgosti. As fumaroles are in general related to former volcanic activity, the area was mapped intensely; however, no evidence of volcanic rocks or pyroclastic depositions were found. The geothermal anomaly observed here is most likely related to tectonic activity.

Figure 3.14.: A: Pliocene conglomerates **P** of the river Cerava near Sveti Naum monastery (view towards south) which are overlain unconformably by Pleistocene conglomerates **Q**.
B: Galicica mountain front with stepped fault scarps **S** and a "wind gap". In the foreground, carbonate-cemented colluvial breccias **B** can be seen (view towards east).
C: Active normal fault with a dragged palaeosol **P** near Ohrid (view towards east).
D: "Duvalo" hydrothermal field near Kosel, note completely altered and sulfur-impregnated phyllites (view towards east). From Hoffmann et al. (2010).

3.3.4. Hydrology

Lake Ohrid with its simple bathtub-shaped basin covers an area of 358 km^2 with an extent of c. 30 km N-S and c. 15 km E-W. The basin has a maximum water depth of about 290 m, and a total depth of roughly 1,000 m below the present lake level and, therefore, preserves c. 700 m of sediments (Lindhorst et al., 2010; Reicherter et al., 2011). The present lake level of Lake Ohrid (at 693 m a.s.l.) undergoes annual fluctuations of c. 0.2 m. After Stojardinovic (1969) natural lake level shifts are in the range of c. 2 m. It

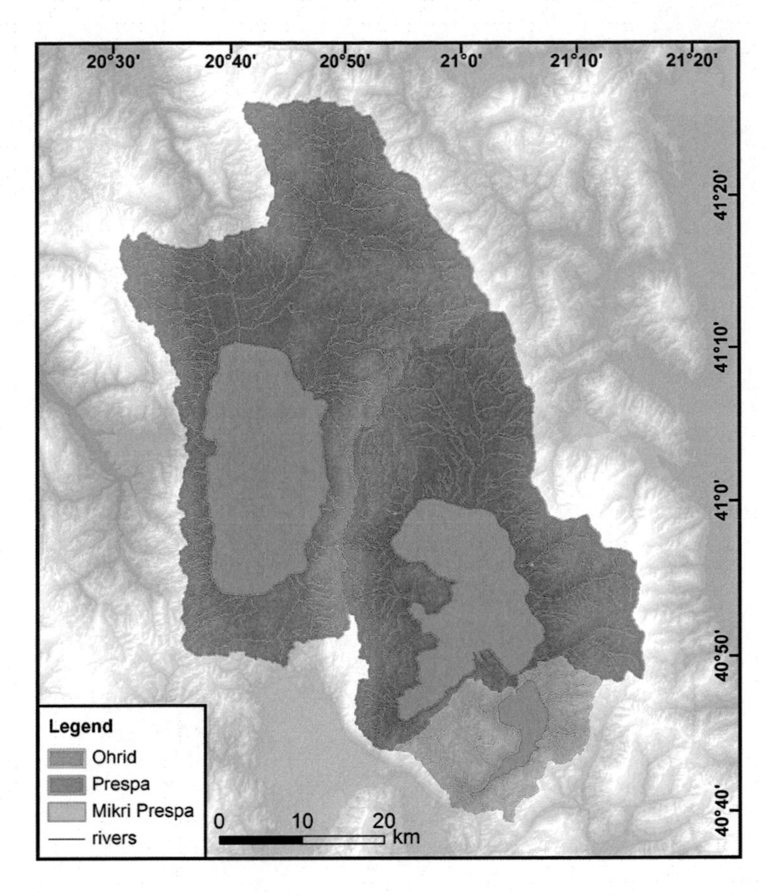

Figure 3.15.: Watersheds and tributary rivers for the Lakes Ohrid Prespa and Mikri Prespa.

reaches its peak value in June and its minimum water level in October/November, while precipitation shows the opposite trend, with a maximum in November and a minimum in July. The reason is that a high portion of this precipitation is snow accumulating in the mountain regions until it starts melting in April (Popovska and Bonacci, 2007). In general, a decrease of 0.67 cm/a for the average water level was observed by Popovska and Bonacci (2007). Climatic conditions are strongly influenced by the water bodies of Lakes Ohrid and Prespa, and by the proximity to the Adriatic Sea, which reduces temperature extremes (Watzin and Naumoski, 2002). An average precipitation for the Lake Ohrid watershed of 907 mm has been determined by Popovska and Bonacci (2007). Temperatures range from c. 26°C in summer to −1°C during winter. Prevailing wind directions are controlled by the basin morphology and are northerly or southerly.

The fresh water lake is presently drained by the Crni Drim River at the northern end of the basin, which accounts for 63% of the water output, while the other 37% are evaporated (Watzin and Naumoski, 2002). It has been suggested that draining had already started in the Lower Pleistocene, as evidenced by deeply incised channels (Dumurdzanov et al., 2005). Today the outflow is regulated to stabilise the water level.

Major fluvial inflows are from the rivers Daljan, Sateska, Cerava and Voljorek (figs. 3.11 and 3.15) with their distinct deltas. Also, high inflow is observed from multiple karstic springs (sublacustrine and subaerial; Albrecht and Wilke, 2009; Popovska and Bonacci, 2007). The water transported by the karstic aquifers originates from Lake Prespa to the east of Lake Ohrid (Matzinger et al., 2006b; Stankovic, 1960). The watershed of

Lake Ohrid extends fairly over 2,393 km^2 including Lake Prespa and its tributaries (see fig. 3.15; Popovska and Bonacci, 2007). Because of the large extent of the karst system and the hydrological connection with Lake Prespa, the exact spatial distribution of the Lake Ohrid drainage basin is hard to determine. The oligothrophic lake is recently endangered by increasing human impact such as tourism, industry, agriculture, etc. expediting the eutrophication process (Matzinger et al., 2006a, 2007; Stankovic, 1960).

4. Palaeostress Analysis

In order to understand the evolution of the basin, the present structural configuration needs to be defined in detail. Following this, estimations of future development can then be made. The influence and variance of external forces leads to a complex system not only in terms of geometry but also in timing; The geodynamic changes which affected the area since the Palaeozoic, are summarised in chapter 3.1 and vary between collision-related compressional stress regimes and back-arc extensional stress regimes.

Due to the cause-and-effect-relationship between strain and stress, fault-slip data measured in the field can be used to determine the past stress states that affected the area. Therefore, populations of fault slip data from striated fault planes were measured to calculate the stress tensor by inversion. For details of methods see chapter 2.1.

The investigated area covers the entire Ohrid Basin extending over c. 1000 km^2 and including the Mountains of Galicica, Mali I Thate and Mokra. Lithologies are mainly composed of limestones and ophiolites, which are, due to their resistance to weathering and their geophysical parameters, able to preserve kinematic indicators. A representative number of suitable fault-slip data (1221 datasets in total) for stress inversion were collected at 45 sites (table. 4.1, fig. 4.1). At three sites (Mali Vlaj, Kosel and Sveti Spas) kinematic indicators were only poorly preserved; these locations were therefore not considered for further analysis. The sites Tushemisht I, Tushemisht II, Tushemisht III, Tushemisht IV and Tushemisht V (Triassic limestones; fig. 4.1) are located at one outcrop along a road and have inhomogeneous strike-slip data. The separation displays many different stress states within each section but with few strike-slip data for each subset. Therefore, the sites at the Tushemisht location are combined to make one outcrop namely Tushemisht I-V. Besides the Tushemisht case, each outcrop is treated separately. Data termed as "rest" are outliers, which could not be explained by any of the found stress states and are too few to make up another stress state. Sites with many outliers such as Kafasan, Ar et Mar, or Pogradec III consist of large inhomogeneous datasets from many but small fault planes creating many stress states made up of only two or three data each. Thus, the amount of outliers for those outcrops is very high.

Besides that, the data were analyzed as independent from the fault size and kinematic indicator quality. An indirect rating of the fault size was reached by measuring a large amount of slip data at bigger outcrops, while at small patches exposing kinematic indicators, only the minimum of five fault slip data were measured. Faults whose sense of slip could not be defined with certainty were not considered at all. Investigations were focused on brittle deformation; however, to get a relative age control on the different stress states, fold axes were measured in addition to intersecting slickensides (see fig. 4.2A).

Stress states are assigned to each dataset depending on the vertical principal axis and on the kinematic classification of faults as reverse, normal or strike-slip (Marrett and Peacock, 1999):

- Vertical σ_1 = tensional
- Vertical σ_2 = strike-slip
- vertical σ_3 = compressional

By analysing misfit angles and the distribution of fault slip data in Mohr circle plots, the data were sorted, evaluated and rated as being created or reactivated by the respective stress regime.

The results of the palaeostress analysis and the final MIM simulation (see chapter 2 for methods) of the subsets of each location are provided in the appendix (figs. A.1 and A.2). From this data stress states were derived by inversion, which are summarised in figure 4.3. In general, one to four subsets were generated for each location with one exception at the Tushemisht I-V site where six subsets were determined. Some of the stress states are very similar (e.g. tensional stresses at Tushemisht I-V, tensional stresses at Galicica II, tensional stresses at Metropol I; fig. 4.3); these data could not be combined to one stress state in the MIM simulation due to the resulting high misfit angles. But for the interpretation the differences between these stresses are so small, that they can be treated as one.

Major tendencies of stress states are: NE-SW compression with a vertical σ_3, NW-SE extension with a vertical σ_1 and E-W extension with a vertical σ_1 accompanied by strike-slip and oblique trends. For exact values for $\sigma_1 \geq \sigma_2 \geq \sigma_3$ of each outcrop see table 4.1. The relative timing was determined by comparing trends of intersecting striations and fold axes (fig. 4.4). This allowed three periods of deformation to be defined.

- orogenic period consisting of NE-SW shortening followed by NE-SW extension
- transtensional phase with NW-SE extension and dextral strike-slip associated with the initial opening of the Ohrid Basin
- E-W extension due to uplift

The results of an intense joint mapping campaign carried out by Fuhrmann (2009); Walter (2009) and Peters (2010) show NE-SW, N-S (NNW-SSE), E-W and NW-SE trends of joints along the entire east coast. In these studies the N-S trending fault and joint system was also classified as the youngest by relative age determination. Orientations are,

Table 4.1. *(facing page)*: Table of palaeostress sites at the Lake Ohrid Basin. The quality of kinematic indicators is expressed by 1=very good, 2=good, 3=poor; the type of kinematic indicators is expressed by 1=mineral crystalisation (cc=calcite, cy=chrysotile), 2=unspecified steps, 3=riedel shears, 4=unspecified scratches. The stratigraphic position of the Korabi serpentinites is not clear as explained in chapter 3.3 and, therefore, no age was assigned to them. K=Korabi units, M=Mirdita units.

	name	latitude	longitude	lithology	rock age	quality	kinematic indicators	no. of subsets		calculated σ_1	calculated σ_2	calculated σ_3	$\Phi = \frac{(\sigma2 - \sigma3)}{(\sigma1 - \sigma3)}$
west coast													
	Ar et Mar	41°00'48.1"	20°38'00.6"	harzburgite (M)	Early to Mid Jurassic	2	2,3, 4	3	subset 1 subset 2 subset 3	267/12 090/15 042/37	147/70 346/40 207/54	001/20 193/45 307/08	0.3 0.1 0.0
	Boces	41°01'92.4"	20°37'56.1"	serpentinite (K)	Early to Mid Jurassic	2	1(cy)	1	subset 1	310/23	066/44	201/36	0.3
	Elen Kamen	41°08'23.6"	20°38'51.4"	limestone (K)	Triassic	1	1(cc), 4	1	subset 1	212/75	013/25	107/04	0.3
	Hudenisht I	40°56'45.2"	20°37'04.1"	limestone (M)	Late Cretaceous	1	1(cc), 4	1	subset 1	287/21	169/34	037/41	0.0
	Hudenisht II	40°56'44.9"	20°37'04.1"	limestone (M)	Late Cretaceous	1-2	1(cc), 4	1	subset 1	050/75	286/16	190/11	0.1
	Hudenisht III	40°56'45.0"	20°37'26.2"	limestone (M)	Late Cretaceous	1	1(cc), 4	3	subset 1 subset 2 subset 3	050/73 054/20 130/50	162/06 300/75 031/03	254/16 178/11 300/40	0.6 1.0 0.0
	Hudenisht IV	40°56'24.9"	20°38'01.0"	serpentinite (M)	Early to Mid Jurassic	2	1(cy)	2	subset 1 subset 2	335/71 099/43	206/15 213/25	111/14 327/36	0.0 0.1
	Hudenisht V	40°56'31.4"	20°38'05.4"	serpentinite (M)	Early to Mid Jurassic	2	1(cy)	1	subset 1	353/71	188/20	096/05	0.7
	Kafasan	41°05'06.6"	20°36'32.4"	limestone (K)	Triassic	1-2	1(cc), 4	3	subset 1 subset 2 subset 3	290/75 152/71 270/36	137/16 051/07 041/35	045/07 317/18 170/14	0.0 0.3 0.1
	Lini I	41°03'07.9"	20°38'41.7"	limestone (K)	Triassic	2	1(cc), 4	4	subset 1 subset 2 subset 3 subset 4	298/76 163/82 267/31 172/51	136/16 045/07 091/64 033/31	049/05 325/06 167/15 291/21	0.0 0.2 0.1 0.5
	Lini II	41°03'33.9'	20°38'03.0"	limestone (K)	Triassic	2	4	2	subset 1 subset 2	275/05 190/43	185/00 321/37	084/85 074/25	1.0 0.3
	Mali Vlaj	41° 07'12.4"	20°37'19.5"	limestone (K)	Mid Jurassic	3	--	0					
	Piskupat	41°01'58.7"	20°37'42.5"	serpentinite (M)	Early to Mid Jurassic	1-2	1(cy)	3	subset 1 subset 2 subset 3	299/01 238/07 275/34	203/74 114/76 126/51	029/16 333/37 005/00	0.2 0.1 0.2
	Pogradec I	40°55'18.6"	20°38'01.8"	limestone (M)	Late Cretaceous	2	4	2	subset 1 subset 2	059/77 058/83	236/15 165/01	149/00 257/06	0.5 0.3
	Pogradec II	40°55'16.7"	20°38'11.3"	limestone (M)	Late Cretaceous	2-3	4	1	subset 1	130/72	283/17	018/07	0.8
	Pogradec III	40°55'24.2"	20°38'03.6"	serpentinite (M)	Early to Mid Jurassic	1	1(cy)	2	subset 1 subset 2	036/78 075/27	118/00 191/40	219/12 321/38	0.5 0.1
	Pogradec IV	40°55'44.1"	20°38'34.9"	limestone (M)	Late Cretaceous	1	1(cc), 4	2	subset 1 subset 2	334/75 143/59	157/16 239/18	068/01 029/14	0.1 0.3
	Radolishta	41°10'58.1"	20°37'28.1"	limestone (K)	Triassic	2	1(cc), 4	3	subset 1 subset 2 subset 3	087/00 055/10 127/07	359/75 305/70 222/71	177/12 150/25 035/19	0.7 0.2 0.3
	Radozda Autocamp	41°05'50.7"	20°37'94.3"	greywacke (K)	Devonian	3	4	2	subset 1 subset 2	071/87 187/21	303/00 065/51	215/03 294/37	0.6 0.2
	Shien Naum I	41°00'07.0"	20°38'05.0"	harzburgite (M)	Early to Mid Jurassic	1	2,3, 4	4	subset 1 subset 2 subset 3 subset 4	161/04 211/71 068/46 236/10	070/14 331/15 207/36 331/27	269/77 064/16 314/22 132/53	0.2 0.0 0.0 0.3
	Shien Naum II	41° 00'13.1"	20°38'01.3"	harzburgite (M)	Early to Mid Jurassic	2	4	3	subset 1 subset 2 subset 3	158/03 075/73 076/33	067/13 308/10 339/09	262/78 217/13 235/55	0.7 0.5 0.4
	Shien Naum III	41°01'36.9"	20°38'13.1"	limestone (M)	Late Cretaceous	2	2,3, 4	3	subset 1 subset 2 subset 3	089/75 039/20 106/16	209/08 165/50 358/51	301/13 297/30 204/26	0.2 0.0 0.2
	Sv. Arhangel I	41°06'25.0"	20°37'57.0"	limestone (K)	Triassic	2	1(cc), 4	1	subset 1	107/75	353/09	258/14	0.3
	Sv. Arhangel II	41°06'42.0"	20°38'00.6"	limestone (K)	Triassic	3	4	3	subset 1 subset 2 subset 3	279/13 271/81 302/53	010/04 024/04 175/25	116/76 115/08 073/26	0.2 0.1 0.8
east coast													
	Daljan	41°07'50.3"	20°46'41.8"	limestone (K)	Triassic	1	1(cc), 4	2	subset 1 subset 2	336/12 138/74	241/14 242/03	116/75 331/16	0.4 0.1
	Elsani I	41°02'02.6"	20°48'38.3"	limestone (K)	Triassic	2-3	3, 4	2	subset 1 subset 2	250/41 018/74	152/08 173/13	055/48 265/06	0.6 0.2
	Elsani II	41°02'03.8"	20°48'44.4"	limestone (K)	Triassic	1	4,3	3	subset 1 subset 2 subset 3	321/52 190/86 187/22	168/34 347/02 046/62	069/13 075/02 284/18	0.7 0.3 0.7
	Galicica I	40°57'06.3"	20°48'02.8"	limestone (K)	Triassic	2	1(cc), 4	2	subset 1 subset 2	264/70 086/57	359/03 186/07	090/20 280/32	0.1 0.3
	Galicica II	40°57'28.7"	20°48'20.7"	limestone (K)	Triassic	1	1(cc), 4	3	subset 1 subset 2 subset 3 subset 4	240/70 285/71 235/75 099/71	100/23 021/01 006/17 005/25	000/11 111/19 101/11 274/19	0.2 0.0 0.0 0.2
	Galicica III	40°57'49.3"	20°49'02.6"	limestone (K)	Triassic	3	1(cc), 4	1	subset 1	081/82	206/05	296/07	0.9
	Galicica IV	40°58'13.4'	20°49'32.3'	limestone (K)	Triassic	1	1(cc), 4	1	subset 1	350/77	196/12	105/05	0.3
	Kosel	41°10'33.9"	20°51'44.4"	limestone (K)	Triassic	1	1(cc), 4	0					
	Ljubanista	40°55'07.8"	20°46'24.9"	limestone (K)	Triassic	2	1(cc), 4	1	subset 1	358/69	230/14	135/16	0.2
	Metropol I	41°03'38.1"	20°48'20.9"	limestone (K)	Triassic	1	1(cc), 4	3	subset 1 subset 2 subset 3	259/71 148/81 267/19	016/10 013/21 006/27	109/17 277/06 150/53	0.2 0.5 0.4
	Metropol II	41°03'51.3"	20°48'30.6"	serpentinite (K)	-	1	1(cy)	2	subset 1 subset 2	250/70 245/75	358/05 350/05	091/19 081/15	0.1 0.4
	Metropol III	41°03'51.3"	20°48'19.3"	limestone (K)	Triassic	2	4	1	subset 1	287/72	202/01	109/18	0.1
	Podmolje	41°08'45.6"	20°45'79.1"	limestone (K)	Triassic	2	1(cc), 4	2	subset 1 subset 2	292/71 257/71	110/18 185/05	201/00 276/15	0.1 0.3
	Sveti Petka	41° 07'49.8"	20°50'28.8"	limestone (K)	Triassic	1	1(cc), 4	2	subset 1 subset 2	244/42 332/38	073/58 145/53	337/04 238/05	0.1 0.3
	Sveti Spas	40°59'30.1"	20°48'43.8"	limestone (K)	Triassic	3	4	0					
	Sveti Stefan	41°4'22.5"	20°48'19.1"	limestone (K)	Triassic	3	4	1	subset 1	293/81	133/09	044/03	0.8
	Tushemisht I-V	40°53'59.2" 40°53'57.2" 40°53'54.4" 40°53'52.8" 40°53'43.5"	20°43'27.0" 20°43'26.7" 20°43'22.1" 20°43'20.8" 20°43'27.3"	limestone (K) limestone (K) limestone (K) limestone (K) limestone (K)	Triassic Triassic Triassic Triassic Triassic	2 1 1 1 1	1(cc), 4 1(cc), 4 1(cc), 4 1(cc), 4 1(cc,mn), 4	6	subset 1 subset 2 subset 3 subset 4 subset 5 subset 6	075/86 051/70 134/70 153/03 084/29 320/38	183/02 199/22 336/30 249/73 242/60 153/52	273/04 293/10 240/06 062/17 349/10 055/06	0.1 0.1 0.2 0.3 0.2 0.5

however, ± N-S directed and these account for 84% of the joint directions (Fuhrmann, 2009). In addition, by separating data into homogeneous subsets it was shown that the coast is highly segmented. Thus, within one block only the NW-SE and NE-SW trending joints occur, while on the edges of a block, which are affected by the recent tectonic stress and normal faulting processes, an additional N-S trending set of joints is observed.

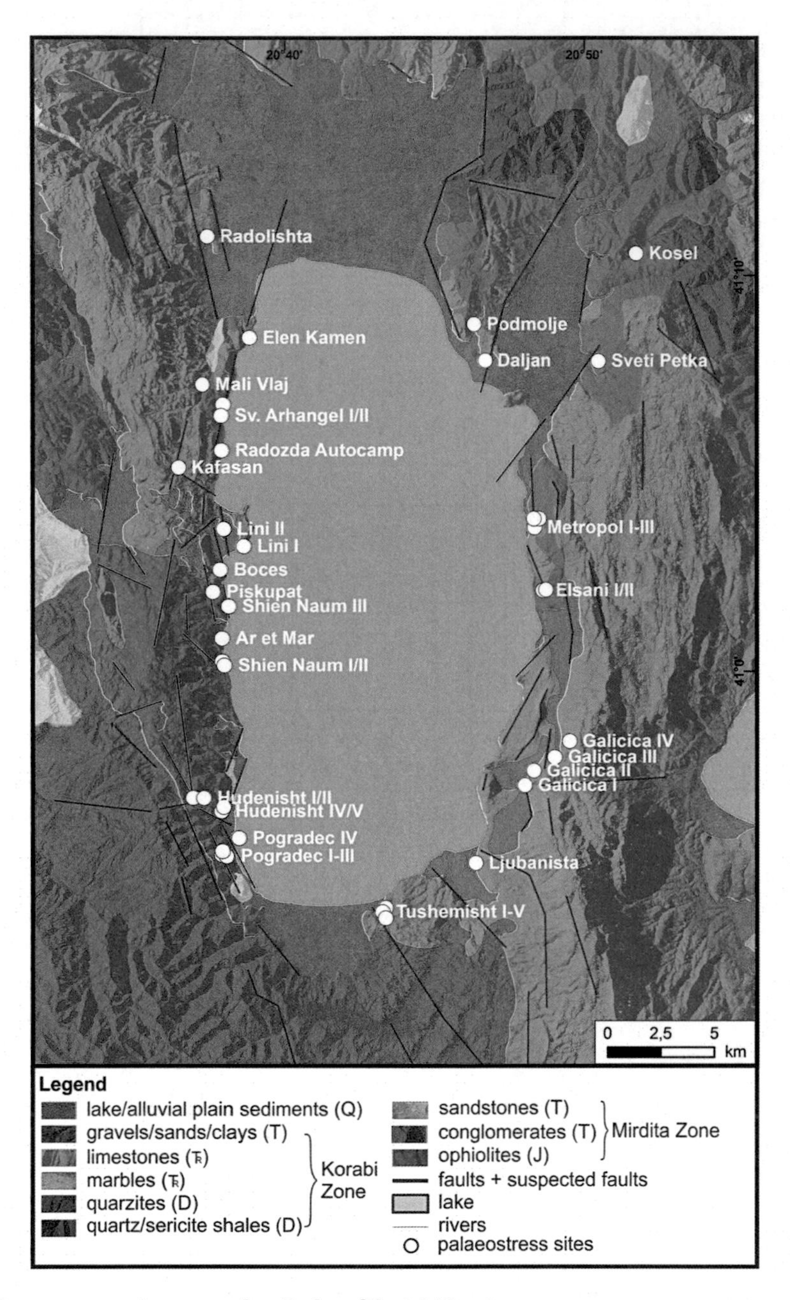

Figure 4.1.: Palaeostress sites at the Lake Ohrid Basin.

4.1. Orogenic Phase

The first phase of deformation is termed the "orogenic phase" as the stress field is dominated by collision tectonics of the Alpine orogenesis. In the palaeostress data, it is characterised by compressional stresses trending mainly NE-SW followed by a phase with

NE-SW trending extensional stress accompanied by few strike-slip stress states.

Besides two sites, ductile compressional stresses preserved in fold data are NE-SW directed (figs. 4.6A and 4.4). In contrast, fold data analysed by Peters (2010) show a NW-SE compressional trend along Ohrid Hill (see figs. 4.6A and 4.4), which corresponds to the observed NW-SE trending stress at the nearby Daljan location. The stress field of the compressional phase is therefore dominated by the NE-SW compression which is also observed in the large thrusts visible in the NW-SE striking suture zone between the Korabi and Mirdita units; where the Korabi zone is thrust onto the Mirdita Ophiolites (for details see chapter 3.3). Hence, the NW-SE trending compression is interpreted as local phenomena occurring only at two locations in the northeast of the lake, most probably caused by inhomogeneous blocks positioned at a lateral ramp. Evidence for large displacements can be seen at the Metropol III location. Here, recumbent and other folds (figs. 4.4 and 4.5A) are overlying each other as a result of the collision of the Afro-Arabian plate with Eurasia and the closure of the Vardar ocean with the establishment of N- or NE-directed subduction (Dilek, 2006; Pamić et al., 1998).

The NE-SW compressional regime is followed by a NE-SW trending extensional stress regime (fig. 4.6B) also associated with the orogenic phase. The data show NE-SW extension along NW-SE to E-W trending faults with mainly low to medium stress ratios (between 0.0 and 0.3; only three stress states exceed stress ratios of 0.5) pointing to an overall radial extension associated with uplift of the graben shoulders. Most of the data plot close to the Mohr envelope, which makes the activation of faults by this stress field very plausible. In addition, extensional foliation structures found at Daljan underline the NE-SW extensional trend (white box in figure 4.6B; fig. 4.2C). Strike slip data associated with this orogenic extension show dextral movement along E-W trending faults. This most probably led to the formation of the large E-W trending wind gap between Lake Ohrid and Lake Prespa, which separates the Galicica from the Mali I Thate Mountains (see also chapter 6).

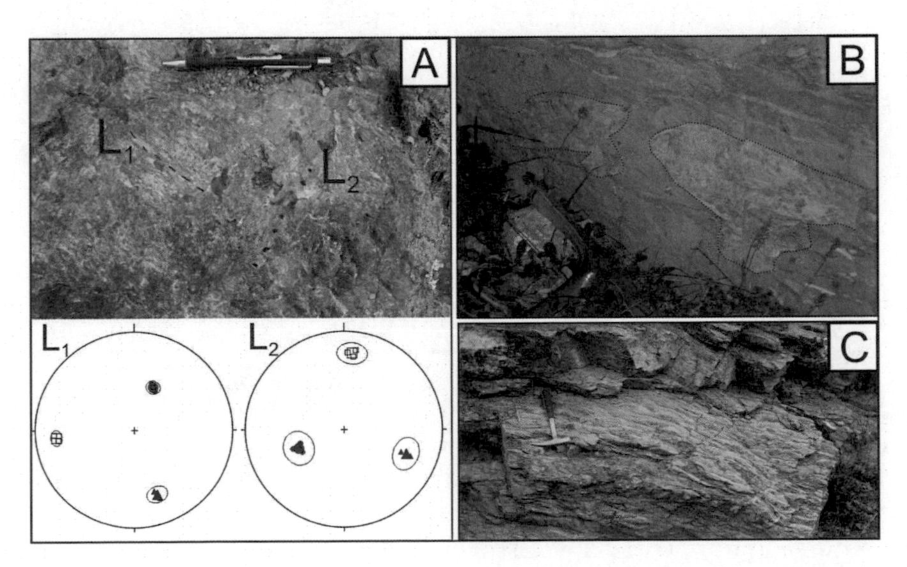

Figure 4.2.: Field photographs of kinematic indicators. A: Intersecting striation with associated PTB-plots at Metropol II demonstrating relative age control. Photograph by Max Arndt. B: Boudinage at the Daljan site. C: Foliation at the top of bedding at Daljan site referring to NE-SW extensional forces.

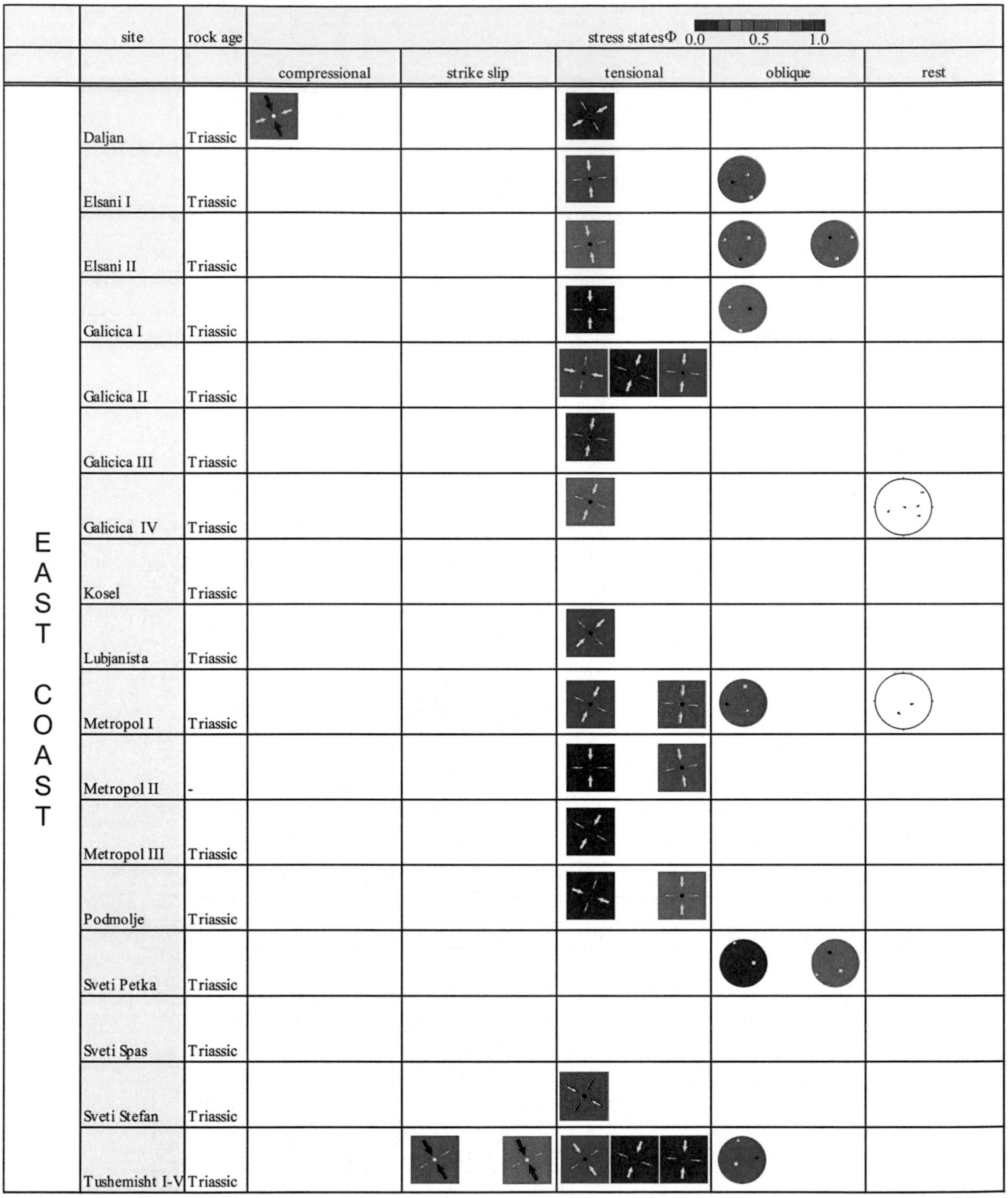

	site	rock age	stress states Φ 0.0 0.5 1.0				
			compressional	strike slip	tensional	oblique	rest
EAST COAST	Daljan	Triassic					
	Elsani I	Triassic					
	Elsani II	Triassic					
	Galicica I	Triassic					
	Galicica II	Triassic					
	Galicica III	Triassic					
	Galicica IV	Triassic					
	Kosel	Triassic					
	Lubjanista	Triassic					
	Metropol I	Triassic					
	Metropol II	-					
	Metropol III	Triassic					
	Podmolje	Triassic					
	Sveti Petka	Triassic					
	Sveti Spas	Triassic					
	Sveti Stefan	Triassic					
	Tushemisht I-V	Triassic					

Figure 4.3.: Stress states obtained for each outcrop showing the stress ratio Φ, the three principal stresses and the rest of each dataset. The colour code for Φ is given in the first line. The column rest shows all data, which could not be assigned to any of the stress states in a tangent-lineation diagram. Each arrow represents the pole to the fault plane and the slip direction of the footwall block.

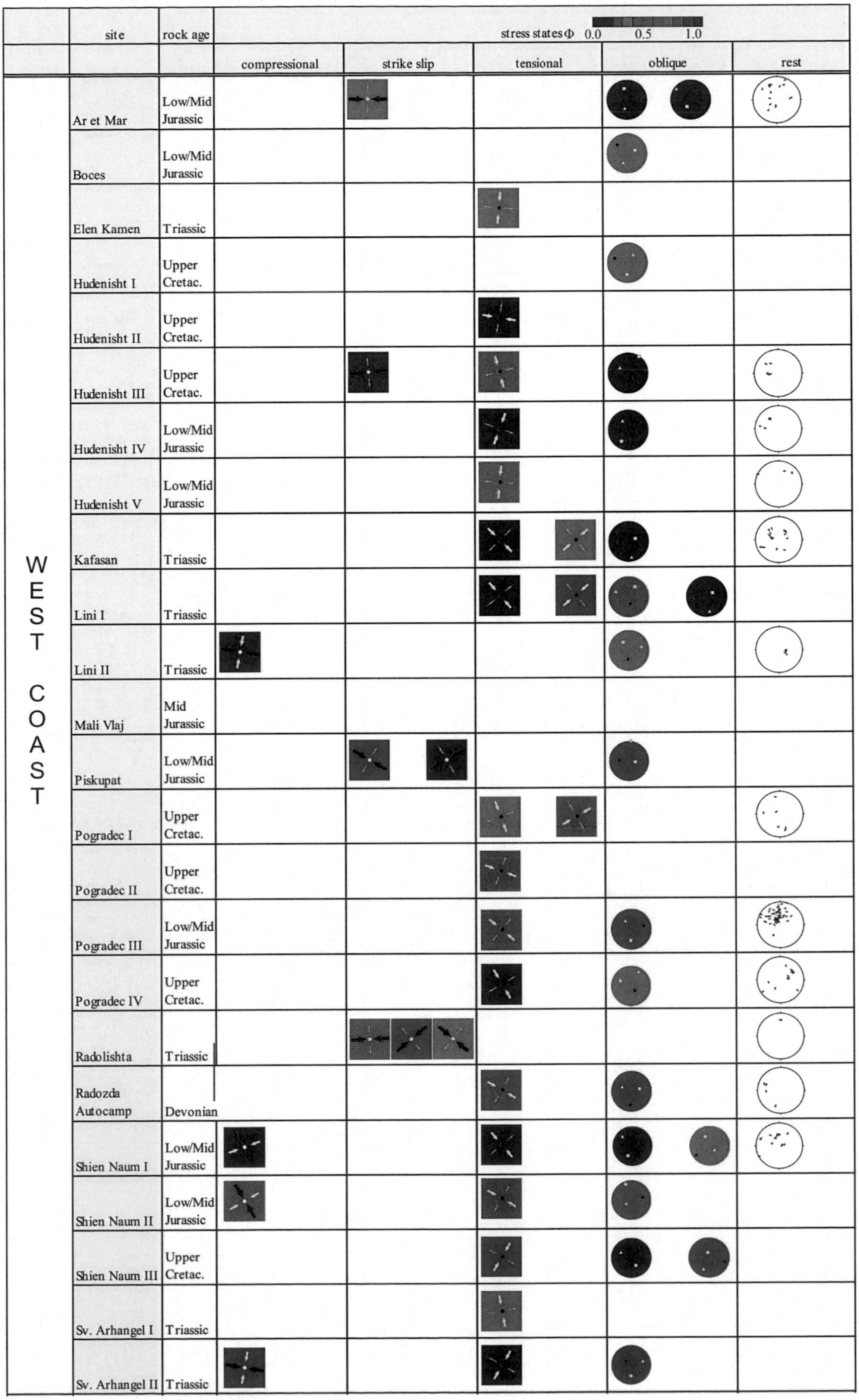

	site	rock age	stress states Φ 0.0 0.5 1.0				
			compressional	strike slip	tensional	oblique	rest
WEST COAST	Ar et Mar	Low/Mid Jurassic					
	Boces	Low/Mid Jurassic					
	Elen Kamen	Triassic					
	Hudenisht I	Upper Cretac.					
	Hudenisht II	Upper Cretac.					
	Hudenisht III	Upper Cretac.					
	Hudenisht IV	Low/Mid Jurassic					
	Hudenisht V	Low/Mid Jurassic					
	Kafasan	Triassic					
	Lini I	Triassic					
	Lini II	Triassic					
	Mali Vlaj	Mid Jurassic					
	Piskupat	Low/Mid Jurassic					
	Pogradec I	Upper Cretac.					
	Pogradec II	Upper Cretac.					
	Pogradec III	Low/Mid Jurassic					
	Pogradec IV	Upper Cretac.					
	Radolishta	Triassic					
	Radozda Autocamp	Devonian					
	Shien Naum I	Low/Mid Jurassic					
	Shien Naum II	Low/Mid Jurassic					
	Shien Naum III	Upper Cretac.					
	Sv. Arhangel I	Triassic					
	Sv. Arhangel II	Triassic					

Figure 4.3.: (continued)

The brittle compressional and strike-slip indicators are primarily distributed along the west coast, and only found at a few locations on the east coast.

This second period of the orogenic phase is linked to either the initiation of the roll back at the trench or spreading within the magmatic arc, which started in Eocene/Oligocene in Eastern Macedonia and migrated westwards (see chapter 3.1 Dumurdzanov et al., 2005). This period most probably accounts for the wide range of the oblique stresses found in the basin.

During Oligocene to Miocene times the arrival of the Kruja fragment and the slowdown of subduction (see also chapter 2) results in a short-lived period of compression, probably resulting in reactivation of faults and left lateral strike-slip movements along N-S trending faults. The spatial distribution of associated stress states is presented in figure 4.6A.

It was not possible to assign the stress states to each of the compressional periods exactly as the kinematic indicators are all preserved in the same lithology (Triassic limestones) and, therefore, no palaeostress-related stratigraphy can be undertaken. Additionally, the direction of compression is the same, so that changes in directions also cannot be used for age determination.

The orogenic phase can be summarised according to the three successive periods: (1) a significant period dominated by NE-SW compression and the closure of the Vardar ocean. This led to the development of thrust-related and strike-slip faults and folds associated with the NW-SE striking thrust belts, and is evidenced by the present day NW-SE striking lithology; (2) a period of NE-SW extension due to the roll back of the subducted slab; and (3) a short lived period of NE-SW directed compression caused by the the arrival of the Kruja fragment in the subduction zone.

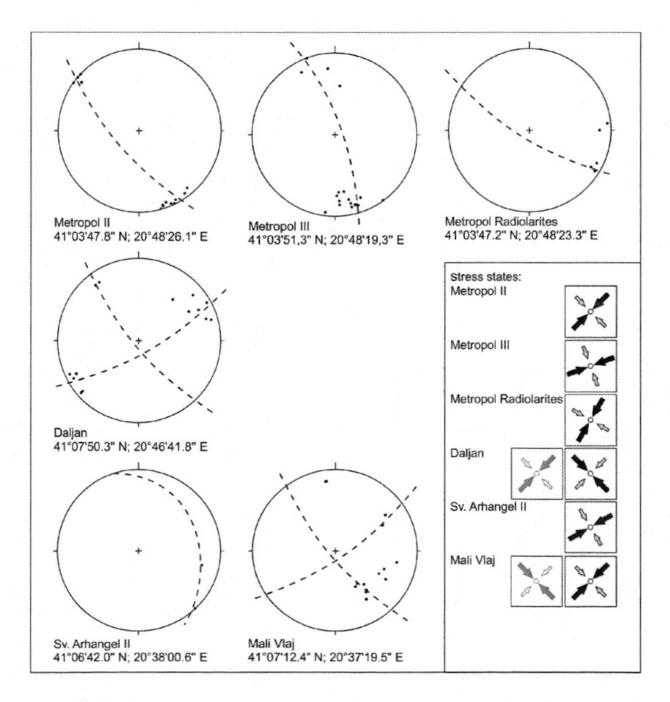

Figure 4.4.: Fold axes and derived stress states. Locations in figures 4.1 and 4.6.

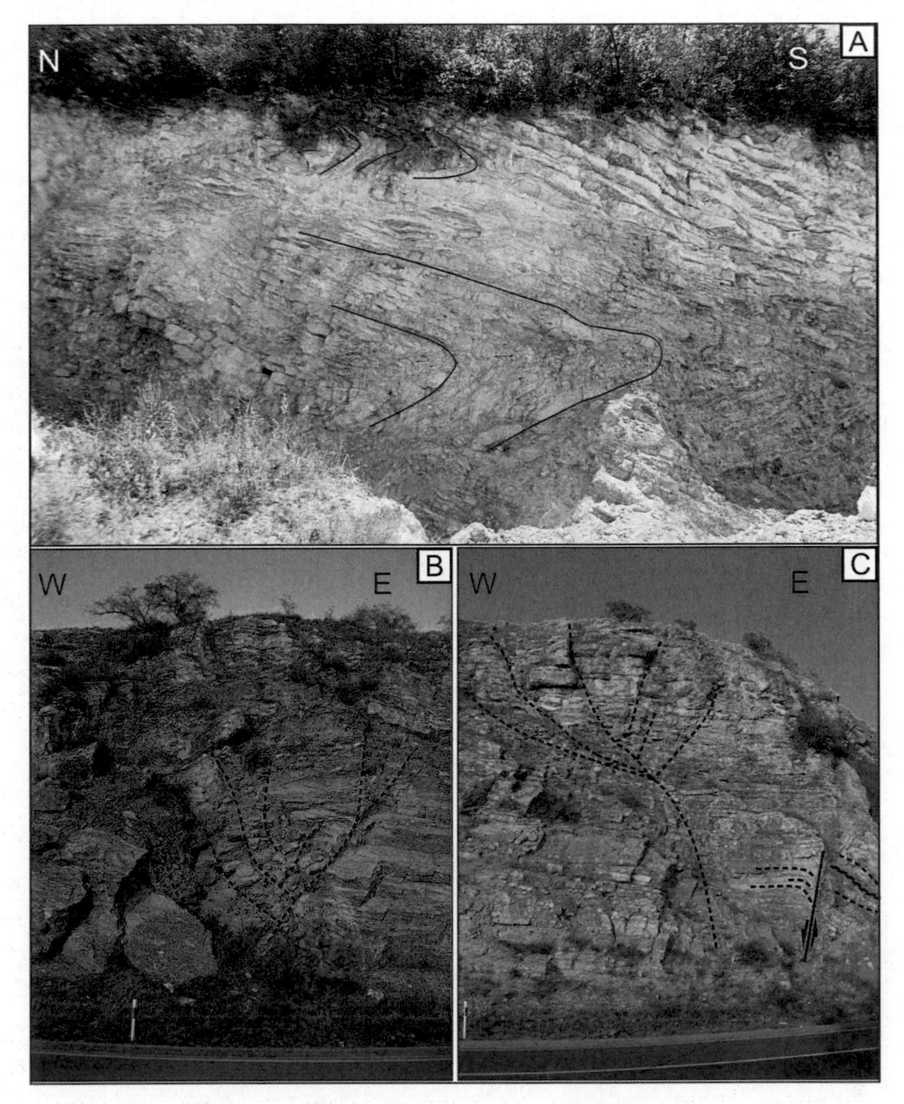

Figure 4.5.: Field photographs of palaeostress sites. A: Recumbent fold at Metropol III. Photograph by Klaus Reicherter. B: and C: Positive flower structures at the Daljan outcrop. (B: western and C: eastern part of the outcrop). Photographs by Tomas Fernández-Steeger. C: To the right a possibly dextral strike-slip fault with a normal component is displayed. It corresponds to the direction of a large NE-SW trending dextral fault located nearby.

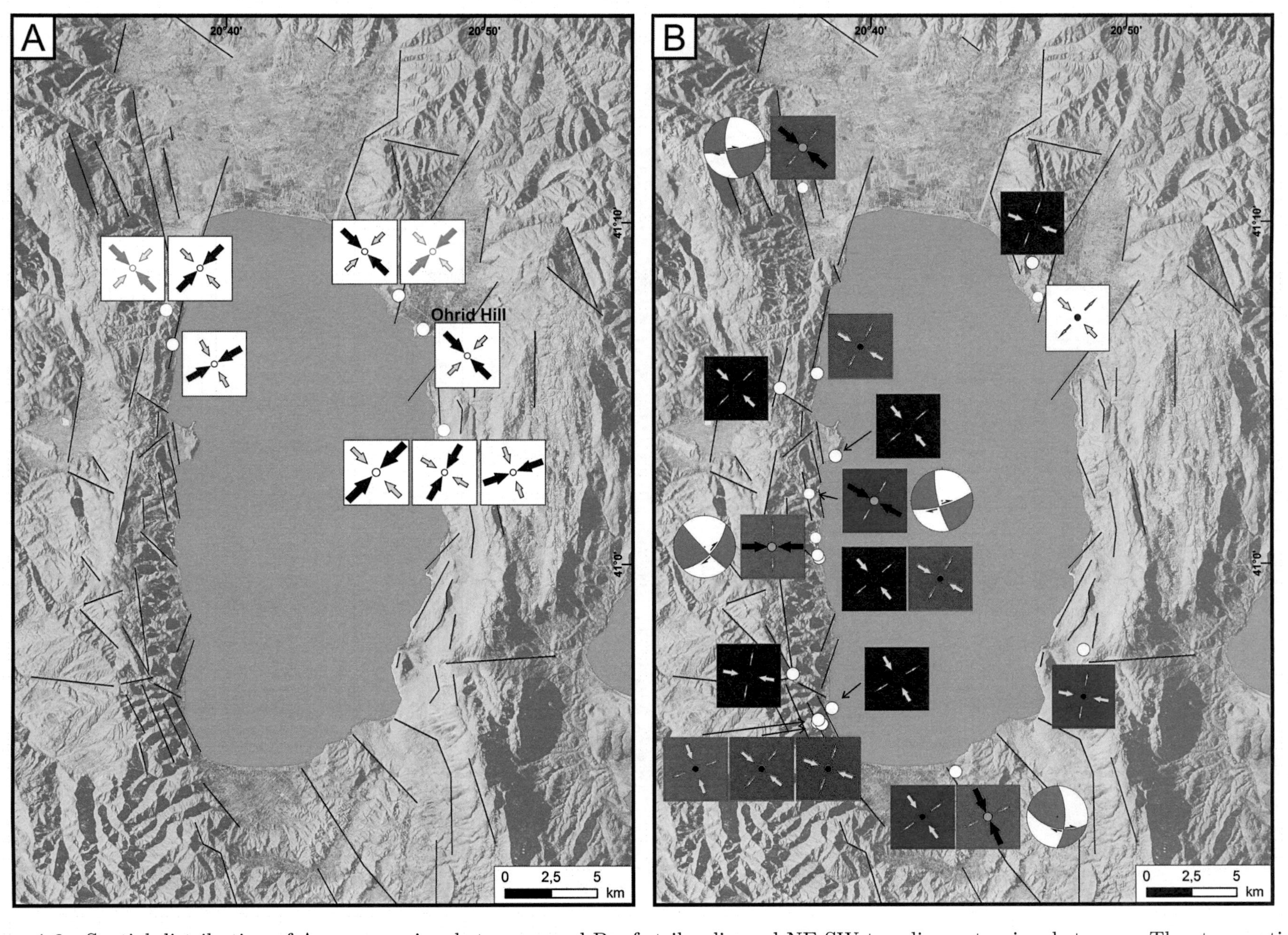

Figure 4.6.: Spatial distribution of A: compressional stresses, and B: of strike-slip and NE-SW trending extensional stresses. The stress ratio Φ of stress states derived from brittle deformation is colour coded see figures 2.2 and 4.3 for comparison; white boxes show data derived from fold axes, grey shaded data in white boxes show minor component at one location.

4.2. Transtensional Phase

Following the compressional orogenic phase, strike-slip faults most likely developed parallel to the orogenic front initiating basin formation. The phase following the orogenic processes results in the pull-apart-like opening of the Ohrid Basin. This transtensional phase is characterised by three modes of stresses which are:

- NW-SE extensional stresses (fig. 4.7A)
- local NW-SE brittle compressional deformation (fig. 4.7B)
- strike-slip faulting along ± NNE-SSW trending dextral strike-slip faults (fig. 4.7B).

From intersecting striations at the Galicica I and Galicica II outcrops, the relative age of NW-SE extensional stresses, postdating the NE-SW extensional deformation of the orogenic phase, is derived. These stresses were observed along NE-SW trending faults which were part of the developing fault system during the pull-apart-like opening of the basin.

While NW-SE extension is mainly observed in outcrops at the central western and eastern coasts, and in only few outcrops towards the southern edges of the lake (fig. 4.7A), the strike-slip component of this phase is only distributed towards the southern and northern ends of the lake (figs. fig. 4.7B and 4.9C). This implies that the straightforward NW-SE extension is mainly located in the centre, while at the edges the regime changes to a strike-slip regime testifying to the duplex shape of a pull apart basin (see fig. 4.9C). This is supported by data from Lindhorst et al. (2012b), who classified the basin as a "symmetrical graben [...] with master normal faults on either side of the pull apart basin".

Evidence of compressional brittle deformation was only found along NW-SE trending faults located at the west coast and at the Daljan outcrop, where two large positive flower structures conserve the strike-slip related compressional stress (fig. 4.5B, C); these flower structures at Daljan are classified as local phenomena at step overs or compressional bends. The nearby coastline bends in a manner, which leads under the influence of a subvertical, dextral strike-slip fault to local compression.

Concerning the NW-SE extensional stresses of the transtensional phase, the stress ratio is low with values generally between 0.0 and 0.3 indicating mainly radial extension. Mohr circle plots show mostly newly fractured faults with few reactivations evidenced by points plotting close to the Mohr envelope. Compressional stresses have a wide variety of Φ ranging from 0.1 to 1.0 indicating a wide range of deformation from radial compression through pure compression to transpression.

The concept of a master fault and the initial opening of the basin is illustrated in figure 4.9A-C. For the evolution of an extensional duplex along a x-axis, a reflected s-shaped fault is assumed. This would open a basin under the influence of a dextral strike-slip movement. Various hints for dextral strike-slip movement in the basin can be found. Palaeostress data suggest dextral strike-slip along ± NNW-SSE trending faults Also the large wind gap results from the later part of the orogenic phase does not have an obvious counterpart at the west coast.

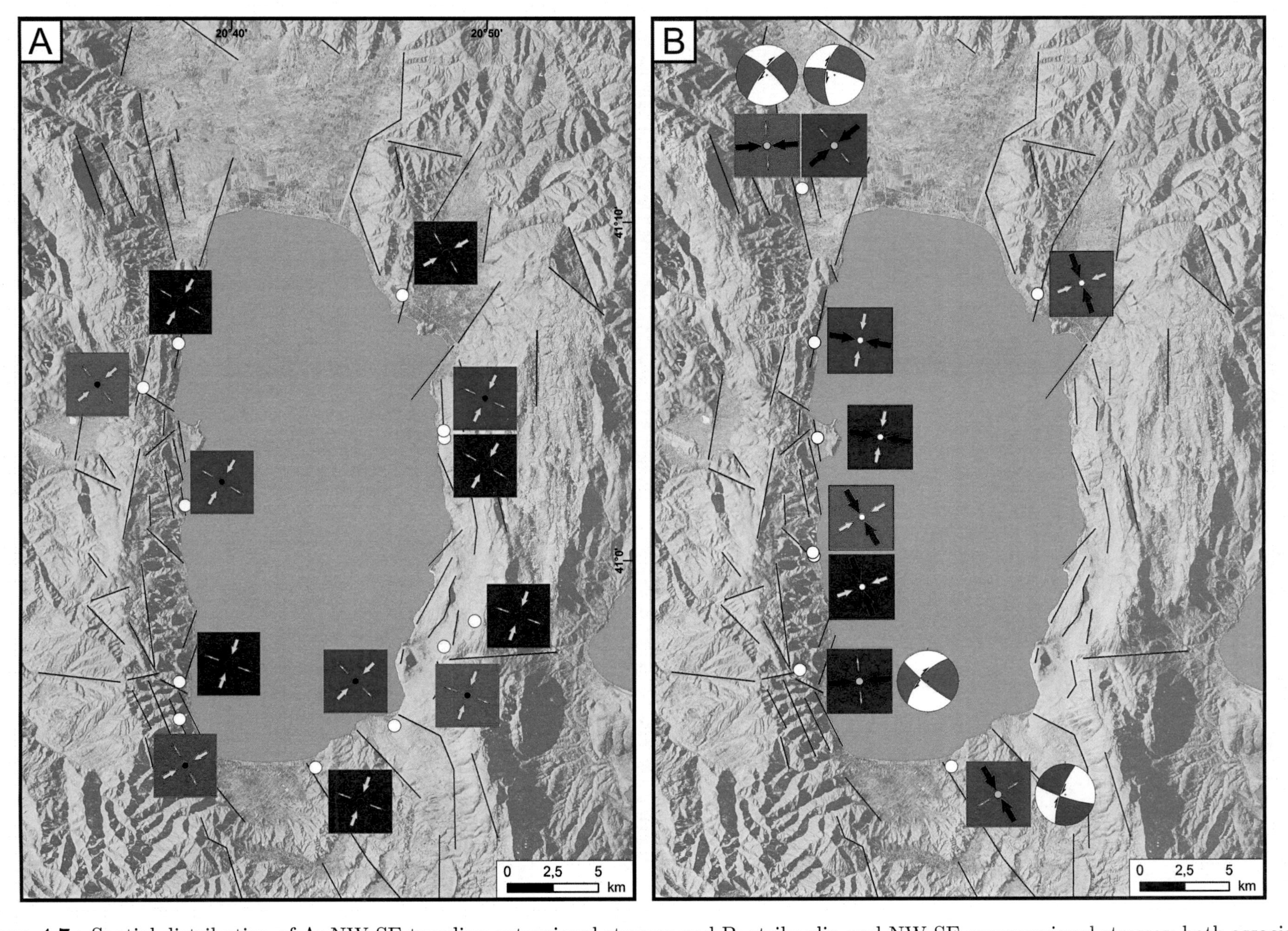

Figure 4.7.: Spatial distribution of A: NW-SE trending extensional stresses and B: strike-slip and NW-SE compressional stresses; both associated with the transtensional phase. Fault plane solutions derived from palaeostress data illustrate activated slip-planes and direction of slip for strike-slip stresses.

A possible match could be one of two E-W trending lineaments west of the Lini Peninsula extending towards Perrenjas. This means in turn a dextral displacement of the E-W trending fault along a large strike-slip fault. Also, Lindhorst et al. (2012b) proposed a dextral master fault according to modeling of Wu et al. (2009). Lindhorst et al. (2012) showed that the faults from within the basin can be extended on land and the authors depict the diamond shaped inner pattern of the initial pull apart basin. From these four main faults, prevailing extension widens the basin while activating new outer faults and becoming more complex. There is evidence that in the south of the basin a wider zone of dextral displacement evolved along at least three faults originating from only one fault (fig. 4.9C). In addition, the duplex system is mirrored in the activity of the so called "Lini" fault (see Lindhorst et al., 2012b), which is active even recently. Figure 4.9C also shows selected fault plane solutions, which support the shown fault traces and kinematics as well as their sense of movement.

The two southern strike-slip data show an activation along NE-SW trending faults, which was derived from strongly deformed outcrops with only small patches conserving kinematic indicators. When the fault plane solutions are compared to the major trends of larger faults, the NW-SE dipping plane is the more prominent direction in the field. Thus, the data measured in the field reflect the true stress field, but were actually derived from minor antithetic faults.

4.3. Extensional Phase

The change of the stress field in Late Miocene/Quaternary times led to overall extension with predominant normal faulting and associated basin widening. This setting partly reactivated faults during E-W directed extension, and more importantly generated the impressive N-S trending normal faults, which today form the most prominent tectonic features in the landscape of the Lake Ohrid Basin. E-W extensional stresses along ± N-S trending normal faults are distributed along both the eastern and western coasts (fig. 4.8). Stress ratios are very inhomogeneous and vary between 0.1 and 0.9 (fig. 4.8). Especially in the southern half of the basin they reach high values. Low values resemble radial extension and therefore are related to extensional forces during uplift, whereas values between 0.25 and 0.75 indicate pure extension. The stress ratio of 0.9 from the Galicica III outcrop is possibly influenced by the nearby E-W trending fault.

There are many N-S trending normal faults but a lot of the vertical stresses are taken up by highly segmented NE-SW and NW-SE trending faults. This leads to a zigzag like coastline especially along the southern and central parts of the east coast. Data from the works of Fuhrmann (2009); Walter (2009) and Peters (2010) show that there is a block-like distribution of faulting leading to certain patterns of fractures, depending on whether data were gathered within or on the edges of a block. This resembles the reactivation along faults that were generated during earlier stages and have been taking up stress during more recent deformation events.

Stronger earthquakes are capable of generating new straight N-S trending normal faults which intersect the predefined zones of weakness. Also, the present-day focal solutions of earthquakes (fig. 3.7) mirror exactly the "pseudo"-fault plane solutions obtained from palaeostress analysis. From the works of Fuhrmann (2009); Walter (2009) and Peters (2010) measurements of fractures and joint mapping also show the same trends.

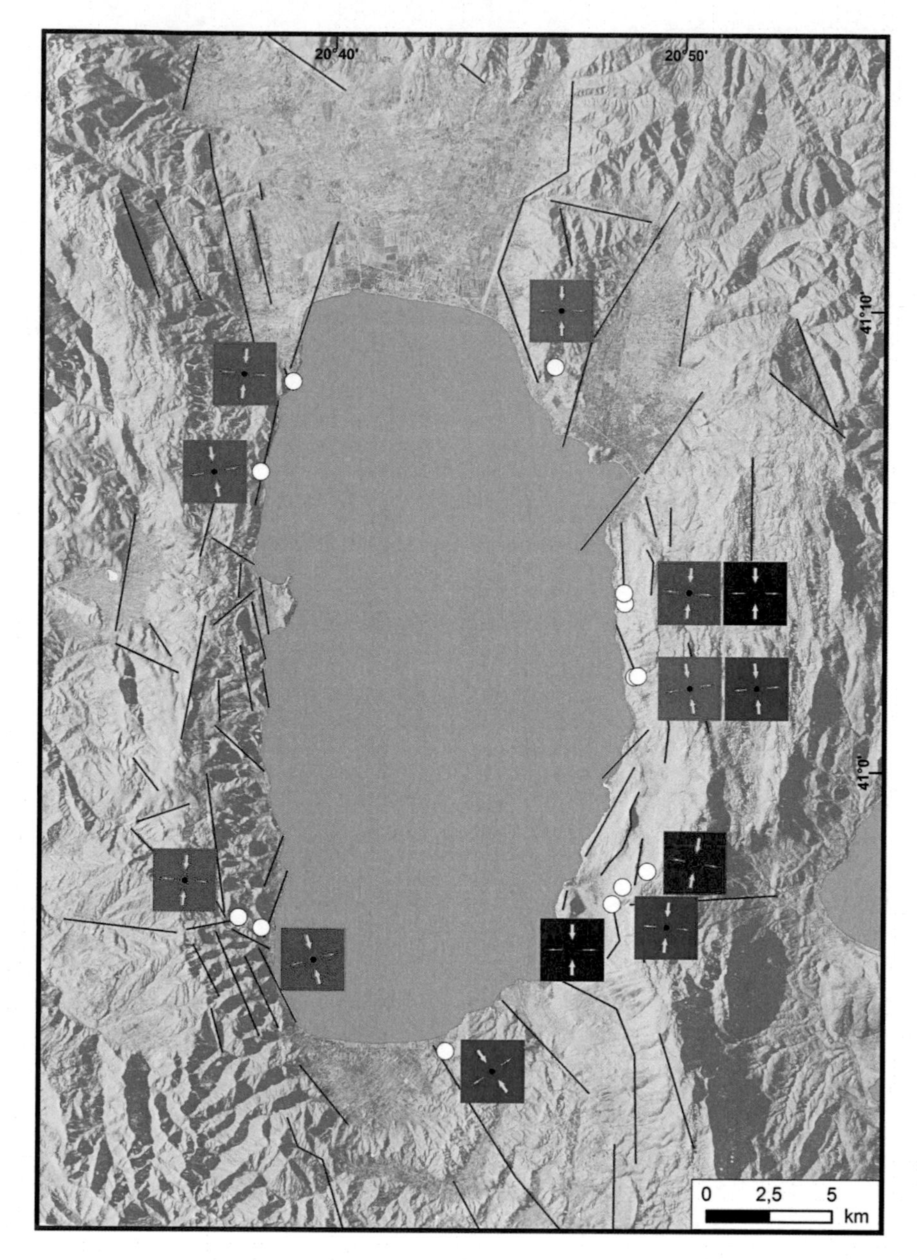

Figure 4.8.: Spatial distribution of E-W trending extensional stresses.

4.4. Discussion

By comparing the relative timing of events to major shifts of the geodynamic setting, the results reflect the following five periods of basin development:

- Orogenic Phase
  - Cretaceous-Paleogene: NE-SW compression
  - Late Eocene-Oligocene: NE-SW extension

  - Oligocene-Miocene: NE-SW shortening and strike-slip movement accompanied by oblique movements

- Mid-Miocene: transtensional phase with NW-SE extension and strike-slip movement accompanied by oblique movements
- Late Miocene to recent: E-W extension

The last two phases are summarised in figure 4.9. The opening of the basin along a releasing bend, where the master fault is shaped like a mirrored "s", in a dextral strike-slip regime is needed to create a pull apart basin. This leads to the development of an extensional duplex, which is characterised by several parallel en-echelon faults taking up the stress and in general keeping up the dextral shear. While the basin is formed there is undoubtedly a normal component involved to a large extent. The building of internal horst and graben structures is inevitable and creates a rather complex system, which is also affected by various overprinting mechanisms. The master fault in figure 4.9A is only a proposed fault, as data from seismics and hydroacoustic surveys could not clarify whether the sets of faults in the centre of the basin are really the oldest or if there is another fault trace hidden beyond sediment cover. According to Lindhorst et al. (2012b), there is no evidence for active faulting in the deep centre of the basin, whereas other signals point to continued widening of the basin.

Lindhorst et al. (2012b) states that the opening of the basin took place along four faults. These are the so called Lini, Pescani faults, Kaneo and Piskupat faults (fig. 4.9). In contrast, the model proposed in this thesis omits the Kaneo fault (WNW-ESE trending fault in the centre of figure 4.9) as a major participating fault and instead uses a fault located further to the northeast called Podmolje fault (fig. 4.9B). From these proposed faults only the Lini and the Podmolje faults are still active today. The Gorenci fault is classified as a sinistral fault in the recent tectonic regime (see Lindhorst et al., 2012b) but has probably formed in the phase of the duplex formation, taking up stress along dextral en echelon faults; it possibly has been reactivated at a later stage changing its deformation to sinistral. The widening of the basin led to a wider deformation zone at the southern tip of the diamond shaped early basin (see figure 4.9C). Pseudo-fault plane solutions show that along the E-W trending main faults the stress regime is mainly NE-SW extensional while towards the northern and southern tips of the basin the stress regimes change to dextral strike-slip.

In the last phase of basin development the stress field changed to pure E-W extension. This is especially seen in the pseudo-fault planes illustrated by figure 4.9D. Most of the extensional stresses also show an oblique component, but orientations are ± N-S directed. Earthquake focal mechanisms (NEIC, 2013) and recent GPS monitoring (Burchfiel et al., 2006) also show recent E-W extension along N- to NNW-striking normal faults.

Many oblique stresses found in the basin can be assigned to one of the Andersonian stress regimes, as only values $\geq 70°$ for $\sigma_{1,2,3}$ were assigned to extensional, strike-slip or compressional stress states. All other were considered to be oblique. This separation leads to the problem, that rotation of blocks, or tilting of strata will cause them to be oblique datasets, although they would normally belong to one of the Andersonian type of faults. On the other hand strike-slip movements are mostly accompanied by minor normal or reverse components resulting, together with rotation of blocks, in an oblique assemblage of stress. Backtilting of blocks was not possible due to a lack of information on the bedding of most of the rocks. Therefore, data from obviously rotated blocks was either not measured at all, or skipped for later analysis. The fault shown in figure 4.5C, which is probably a dextral strike-slip fault, also exhibits significant normal displacement and would thus be classified as oblique. Unfortunately no fault slip data could be recovered from this particular fault. Oblique stresses from several outcrops (Galicica I, Galicica II, Elsani II, Ar et Mar, Pogradec IV, Piskupat, Lini I, Kafasan and Hudenisht IV) are examples of slightly tilted blocks and belong to one of the tensional stresses according to their stress ratio and their orientation of the three principal stresses. The rest of the oblique stress regimes most probably account for the major part to the last part of the orogenic phase or to the transtensional phase.

Looking at the data at different scales, the large scale N-S trending fault system obvious in the landscape morphology is not reflected in the small scale measurements. In contrast to the general trend, the faults along the coastline strike NW-SE or NE-SW, which results in the zigzag line of the coast. This observation is most probably connected to inheritance of faults and the reactivation of the fault system along predefined zones of weakness such as joints, fractures and older faults. However, this pattern also resembles the block model proposed above; the blocks act as seperate units which are displaced against each other and form the curved coastline. Strong segmentation is also described by Lindhorst et al. (2012b) from the eastern part of the lake centre. Here, displacement is taken up by several smaller faults associated with the formation of half grabens and the tilting of blocks, rather than the formation of large faults. A strong tectonic strain affecting the rocks by changing stress regimes is evidenced by strongly fractured rocks and associated micro fractures, the large variety of joint directions, and brecciated fault gouges.

Liermann (2010a) found mudflow deposits incorporating ophiolite fragments in the Kosel area. As no ophiolites are abundant there and only a few serpentinite lenses are present along the entire east coast, it is possible, that during the initial opening of the basin the western graben shoulder was uplifted earlier or to a greater extent. The associated ground movement my have caused a debris flow to occur transporting the material from the west coast eastwards in an early rifting stage. As Lindhorst et al. (2012b) show, it is quite possible that the basin was influenced by a river system in the early rifting stage and therefore the surface transport of debris is a likely process. In contrast, thermochronological dating carried out by Liermann (2010b) suggest that the Prespa-Ohrid shoulder was uplifted before the western shoulder of the Ohrid graben and that the transport direction of molasse sediments was eastward.

A maximum extension rate of c. 10.5 mm/a can be calculated assuming the proposed minimum age of 1.9 Ma (Lindhorst et al., 2012b) and an average basin width of 20 km. The values for the Lake Ohrid Basin are considered reasonable when they are compared to data from other basin and range provinces (Eddington et al., 1987; Thatcher et al., 1999), where overall extension rates between 8-12 mm/a with locally increasing or decreasing values are determined by GPS networks. Even if the activity in the basin is lower recently, there possibly have been times with higher extension rates.

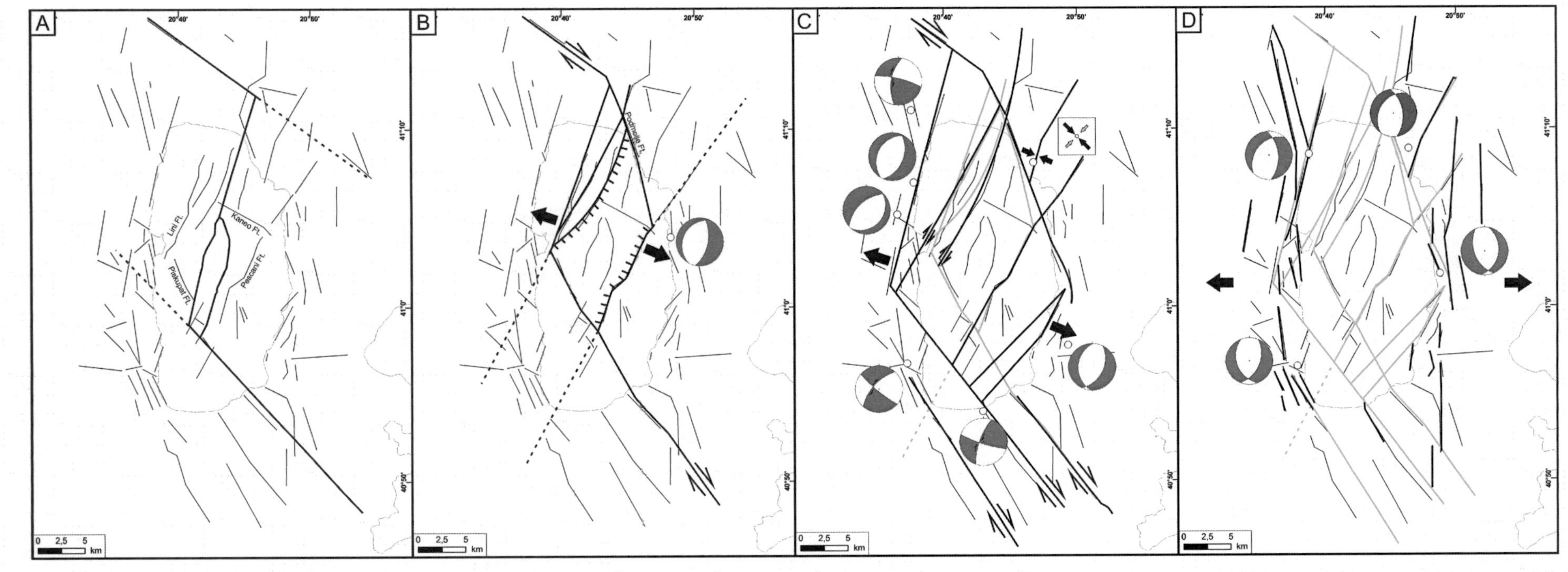

Figure 4.9.: Schematic development of the Lake Ohrid Basin since the beginning of the postcollisional stages. Red lines mark faults derived from field data and maps, fault lines in the lake are adopted from Lindhorst et al. (2012b). All fault plane solutions are derived from palaeostress data. A: Possible trace of the initial fault in the beginning of the transtensional period. B: The bends along the generally NW-SE trending dextral strike-slip fault lead to the transtensional opening of the basin in an extensional duplex. C: Further development of the pull apart basin with exemplary fault plane solutions. D: Shift from the transtensional regime to a pure extensional regime.

5. Sedimentological Investigations

To reconstruct the Holocene shoreline evolution of Lake Ohrid, the coastline was investigated using sediment cores and shallow subsurface geophysics to obtain data on sedimentary and deformational structures (Reicherter et al., 2011). GPR and ERM were applied as non-invasive methods for subsurface investigations. In addition to the geophysical survey, a detailed sediment analysis of short cores was performed; the microfaunal content, the composition of sediments, and their architecture were studied. Macro- and microfossils were sorted and identified for palaeoenvironmental analysis. Magnetic susceptibility, X-ray fluorescence analysis (XRF) and radiocarbon dating were additionally used on cores where applicable (for methods see chapter 2).

Areas of differing sedimentation regimes were mapped. While the plains north and south of the lake are dominated by clastic input related to climate variations and uplift/erosion, the steep western and eastern margins are controlled by recent tectonics. Furthermore, no evidence for a much higher lake level during the Holocene was found in the plains north and south of the lake, except for rare temporary floodings. This is supported by mappings of the limestone cliffs around Lake Ohrid, which yielded no evidence for abrasional platforms or notches as indicators for past high stands.

Five locations were chosen close to the shoreline of Lake Ohrid in order to study sedimentary evidence of past lake level fluctuations and the evolution of the plains (fig. 5.1). At each location radar lines, drilling/coring, and resistivity measurements were applied.

The location "Struga" (fig. 5.2A) is situated east of the town of Struga (41°10′14″N, 20°42′28″E) in the northern plain of the basin near the outlet of Lake Ohrid and close to the river Sateska. Sedimentation here is mainly controlled by alluvial processes and large fan complexes. The former swampy wetlands in the lower areas have been drained for agricultural use.

The second location is south of the town of Ohrid (41°05′38″N, 20°48′18″E) and is situated on the eastern shore of the lake, close to the village Velestovo (fig. 5.2B). Here the fault-related steep slopes of the Galicica Mountains (Hoffmann et al., 2010) generate high relief and a relatively small swampy area with a minor alluvial channel and fan. This site was chosen as it should provide insight into the fault mechanisms driving basin development and give evidence for the interaction between lake level fluctuations and basin extension.

The third site is located in the southern part of Lake Ohrid near the monastery of Sveti Naum (40°55′23″N, 20°45′19″E) and east of the Macedonia/Albanian border in the village of Ljubanista (fig. 5.2C). In this area many subaerial karstic springs feed Lake Ohrid (Albrecht and Wilke, 2009) and the Cerava River discharges into the lake at a delta.

The fourth site is situated at the river mouth of the Daljan River (41°07′17″N, 20°46′33″E) west of the town of Ohrid (fig. 5.2E). A small delta has developed here at the coastline of Lake Ohrid. Together with the Struga and Sveti Naum sites, this area is suitable to study plain evolution.

The Lini Peninsula on the western shore of Lake Ohrid (41°03′18″N, 20°38′15″E) is suitable to study the interaction between faults and lake level behaviour (fig. 5.2D). The peninsula is a flooding platform which is bound by faults to the west and the east (see fig. 5.3). In between the faults, a flat plain has recently developed which is currently being used for agriculture.

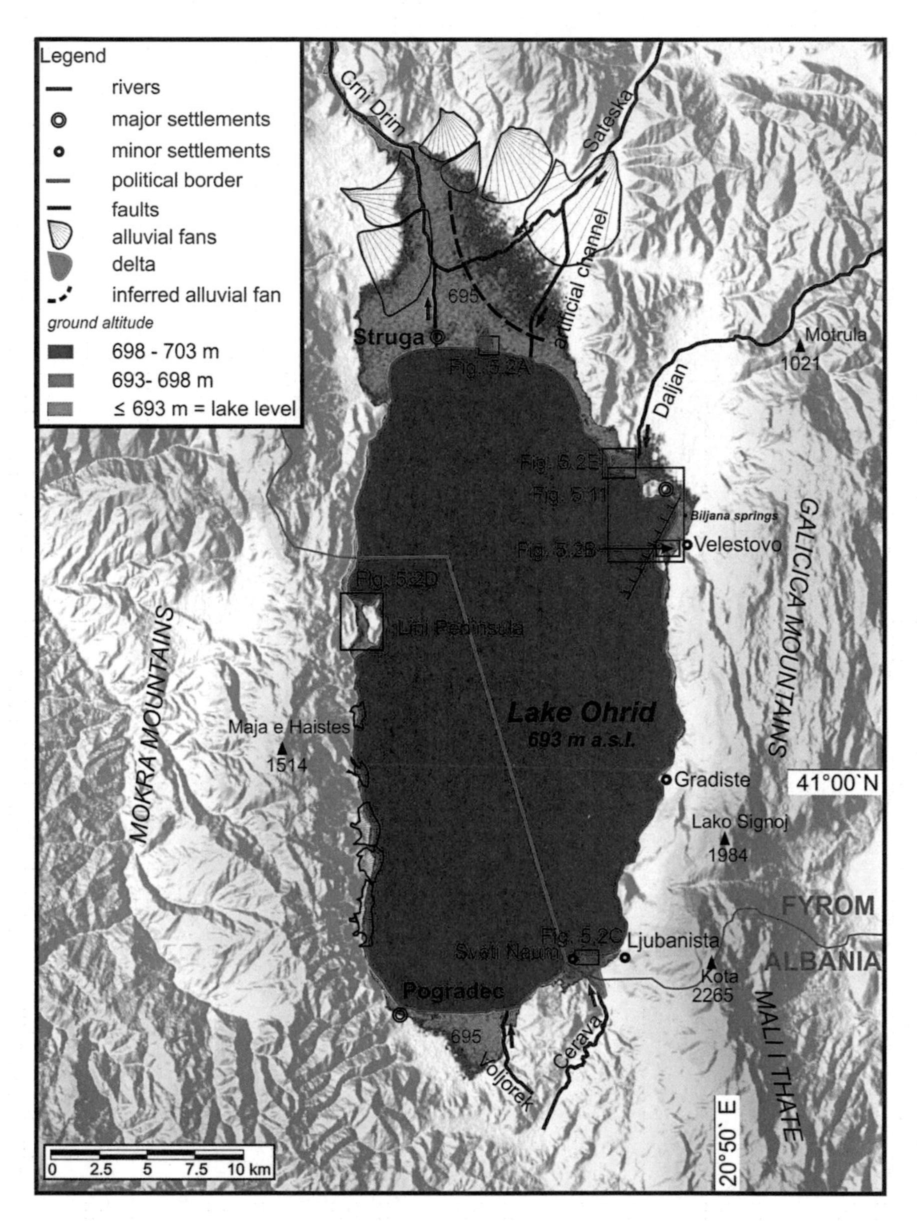

Figure 5.1.: Geographical overview of the Lake Ohrid area in Macedonia and Albania showing major rivers, alluvial fans, plains and deltas. Locations investigated are marked with a box. Modelling of the lake level is based on SRTM data to derive palaeoshorelines. The coloured areas do not exceed heights of 10 m above present day lake level. Note: normal fault off-shore from the Velestovo site. Modified after Hoffmann et al. (2012).

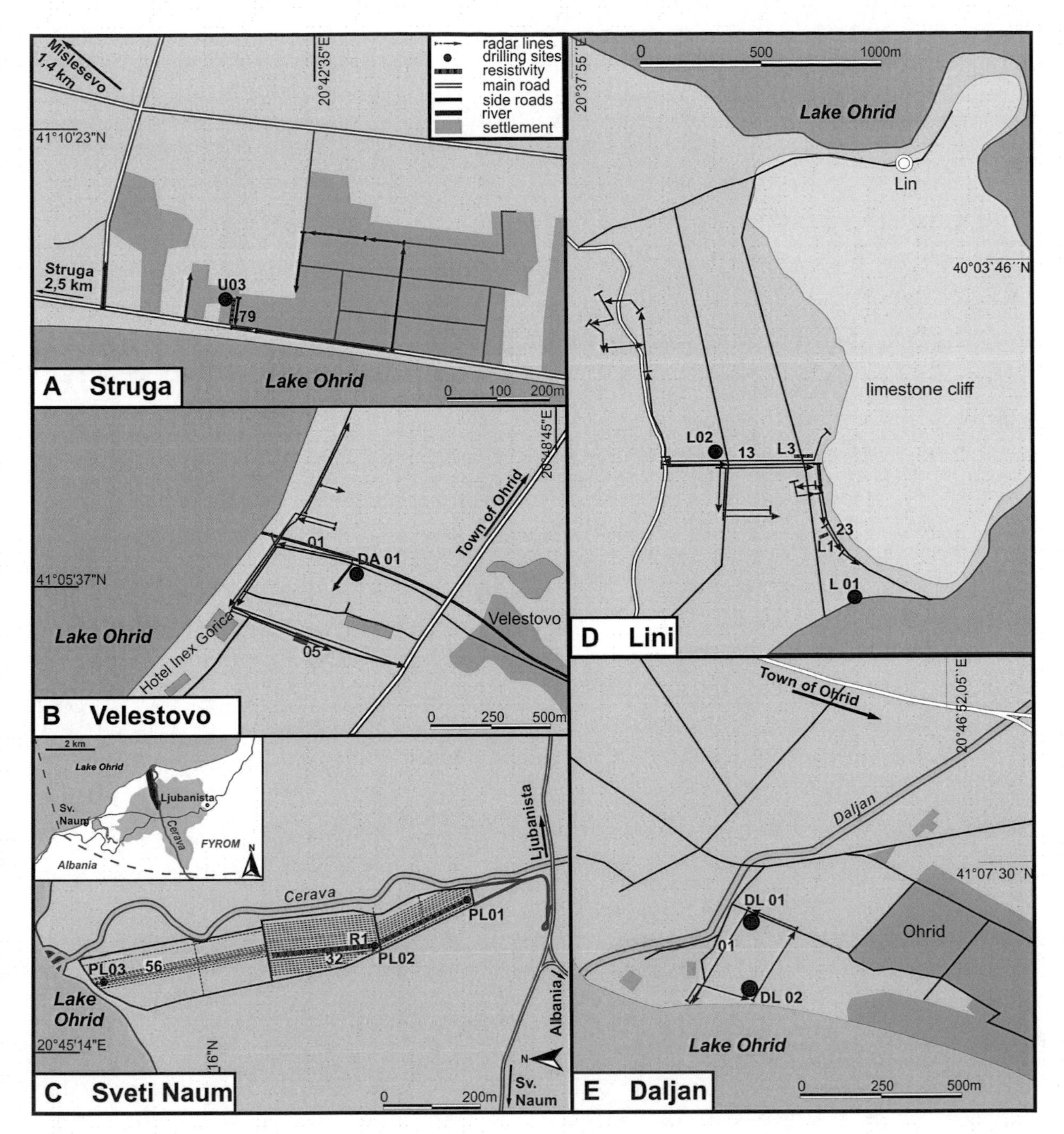

Figure 5.2.: Map locations are shown in figure 5.1. A: Sketch map of the study area near Struga. Electric resistivity measurements were carried out parallel to radar line 79 (see fig. 5.6). U03 marks the drilling site; the core log is illustrated in figure 5.4. B: Radar and drilling localities at the Velestovo site. For core log of DA01 see figure 5.8. The red box shows the segment of GPR line 05 displayed in figure 5.10. C: Sketch map of the investigation site at Sveti Naum/Ljubanista. Inset plan shows the location on Lake Ohrid. Core PL03 is plotted in figure 5.12. Radar reflection lines (no. 56 and 32) are displayed in figures 5.13 and 5.14A. Electric resistivity measurement R1 is illustrated in figure 5.13B). D: Overview of Lini Peninsula measurements. Core logs L01 and L02 are displayed in figures 5.18 and 5.19, radarlines are shown in figures 5.20 and 5.21. Electric resistivity measurement L1and L3 are illustrated in figure 5.22. E: Map of the Daljan delta. For core DL02 see fig. 5.15, the interpretation of radarline 01 is shown in figure 5.17. Modified after Hoffmann et al. (2012).

Figure 5.3.: Interpreted photograph of Lini Peninsula with view to the North. Black lines indicate normal fault traces, red circles show positions of drill cores.

5.1. Struga

5.1.1. Drilling and core logging

The core recovered close to Struga has a total length of 400 cm (fig. 5.4). The water table was encountered at approximately 2.50 m depth which corresponds to the present-day lake level. In general, the lithology of core U03 is differentiated into 4 stratigraphical and sedimentological intervals; the facies indicate deltaic sediments.

(1) 400 - 275 cm: Mixed clastic sequence ranging from rounded gravels to clays. Sieve residues of this interval yield significant lacustrine fossil contents, which mainly consist of remains of charophyte algae (carbonate-coated stems and oogonia), tubes of tubifids, and ostracod valves. The latter taphocoenoses is fairly diverse and contains juveniles and adults (fig. 5.5).

(2) 275 - 175 cm: Deltaic mixed clastic-lacustrine sediments as evidenced by Chara oogonia and ostracod assemblages. Clayey and silty layers are intercalated with fine-grained sands between 275 cm and 233 cm. Above 233 cm, sand and gravel layers with fining-upward sequences are present.

(3) 175 - 80.5 cm: Fluvial sediments made up of sand to fine-grained gravel with fining-upward sequences. The coarser components show an oval shape indicative of a longer fluvial transportation distance.

(4) Anthropogenic filling from 80.5 - 0 cm containing fragments of bricks (at 58.5 cm, 56 cm and 37.5 cm).

The subdivisions are also mirrored in the results of magnetic susceptibility (MS) measurements (fig. 5.4). From the bottom to the top of the core the MS signal generally gets stronger; however, the increase is more dramatic at the identified sedimentary changes. The signal remains almost stable within each of the four sedimentary units. The ostracod fauna in the core material of U03 is diverse and dominated by Candonidae (e.g. Cypria obliqua, C. ophtalmica, Candona parvula, C. lychnitis, C. trapeziformis and C. hart-

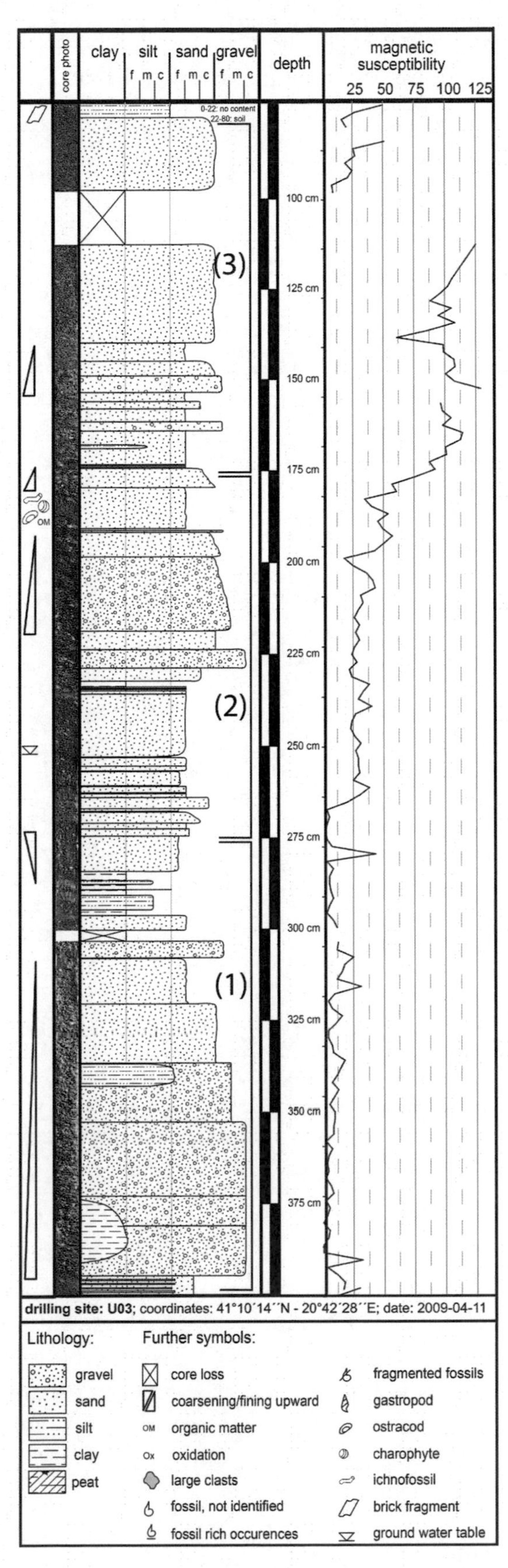

Figure 5.4.: Core log of U03 carried out at the Struga site with a total length of 400 cm. Grain size, fossil content and magnetic susceptibility are displayed. The first 75 cm are not illustrated because of loss of core material. The ground water table can be observed at a depth of about 250 cm by a sudden increase of water content in the sediment. Modified after Hoffmann et al. (2012).

manni; fig. 5.5), associated with Paralimnocythere slavei, P. karamani and Amnicythere (Leptocythere) karamani. Characeae remains and fragmented aquatic shells complement the findings.

5.1.2. Geophysical investigations

A total of 24 GPR profiles have been made in the Struga area (fig. 5.2A). Radar line 79 has a total length of 47 m and depicts the subsurface architecture of the sedimentary intervals (fig. 5.6A). The upper interval, to a maximum depth of 180 cm, is characterised by diffuse horizontal to subhorizontal reflectors caused by diffraction hyperbola of larger clasts. Based on the drilling results, the interval from 0 - 80 cm is ascribed to an anthropogenically influenced layer (e.g. a buried tube at 19.5 m profile distance and the previously mentioned brick fragments), and the following 80 - 175 cm to an interval influenced by fluvial processes. Below this horizon a set of south dipping reflectors up to 300 cm depth were found along the entire length of the profile. The dip angle of the reflections varies between 5° and 10°. Below 300 cm depth (which corresponds to 60 ns) the radar signal diminishes and no interpretable data was obtained. The extinction of electromagnetic waves may be due to the groundwater level (lake level). The resistivity profile (fig. 5.6B) which runs parallel to radar line 79, does not show a comparable high resolution as the GPR. However, it also shows a change around 300 cm depth to very weak resistivity. This could be the result of sedimentary change, but a high water table is the more likely explanation. Additionally, a gentle dip to the south can be observed along the entire profile which might be related to the orientation of the foresets.

5.1.3. Interpretation

The sediment core from the Struga site shows a typical channelised fluvial facies in the upper part. In the lower part, a change into deltaic sediments with fining-up and alternating coarse- and fine-grained layers up to 300 cm depth, including littoral faunal and floral evidence, is recorded. The identified ostracod fauna consists of a rich littoral species assemblage (fig. 5.5). Remains of Chara sp. are also generally found in water depths between 3 and 20 m in Lake Ohrid (Albrecht and Wilke, 2009). Today, Characeae normally occur not deeper than 12 m (see also fig. 5.7). Therefore, the faunal and floral remains in approximately the upper 200 cm of core are interpreted as a lake level high stand. According to the results of GPR and drilling, the lake-ward dipping reflectors are interpreted as delta foresets of the mixed fluvial-lacustrine layer. The shallow topography of the northern Struga plain can be an indicator for a wider extent of the lake during past times. Later, the plain was exposed by lake level drop, or it represents an area of lower subsidence and higher sedimentation rates as evidenced by the large alluvial fans (fig. 5.1).

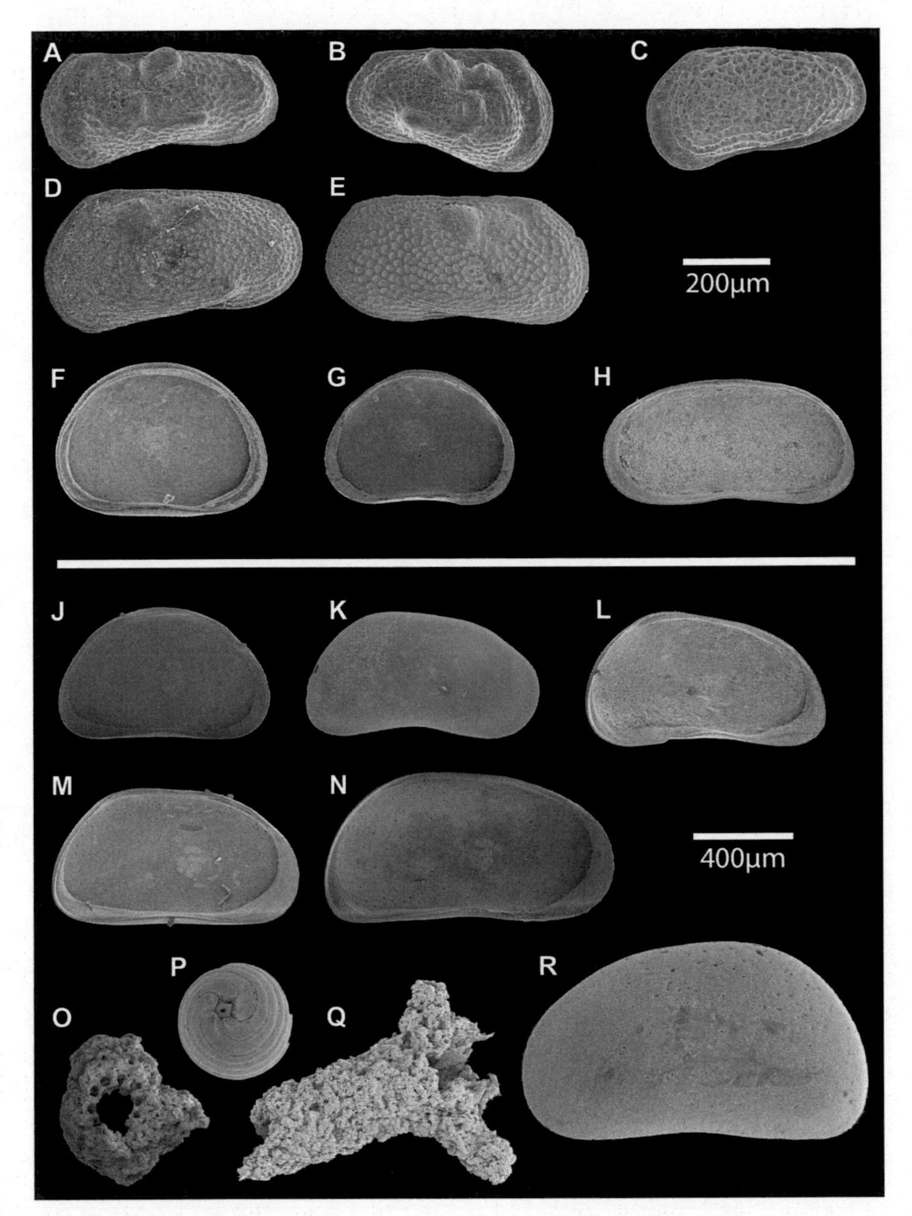

Figure 5.5.: Fossil ostracods samples of various cores. A: *Paralimnocythere karmani* (Petkovski, 1960b) (male), LV, extern., U02 (182-187.5), 560 m; B: *Paralimnocythere karmani* (Petkovski, 1960b) (female), RV, extern., U02 (182-187.5), 480 m; C: *Leptocythere karamani* (Klie, 1939b), LV, extern., U02 (182-187.5), 500 m; D: *Paralimnocythere slavei* (Petkovski, 1969a) (female), LV, extern., DA01 (885-890), 610 m; E: *Paralimnocythere slavei* (Petkovski, 1969a) (male), RV, extern., DA01 (880-885), 610 m; F: *Cypria ophtalmica* (Jurine, 1820), LV, intern., U02 (182-187.5), 435 m; G: *Cypria obliqua* (Klie, 1939a), LV, intern., U02 (182-187.5), 475 m; H: *Candona parvula* (Mikulić, 1961), RV, intern., U02 (182-187.5), 580 m; J: *Candona goricensis* (Mikulić, 1961) (male), LV, intern., DA01 (880-885), 840 m; K: *Candona lychnitis* (Petkovski, 1969b), RV, extern., DA01 (880-885), 985 m; L: *Candona hartmanni* (Petkovski, 1969b), LV, intern., U02 (182-187.5), 1020 m; M: *Candona trapeziformis* (Klie, 1942), LV, intern., U02 (182-187.5), 1000 m; N: *Candona formosa* (Mikulić, 1961), LV, intern., DA01 (880-885), 1240 m; O: Characeae stem, planar, U02 (182-187.5); P: Characeae stem, lateral, U02 (182-187.5); Q: Characeae oogonia, U02 (182-187.5); R: *Candona lychnitis* (Petkovski, 1969b), LV, intern., DA01 (880-885), 1435 m. From Hoffmann et al. (2012).

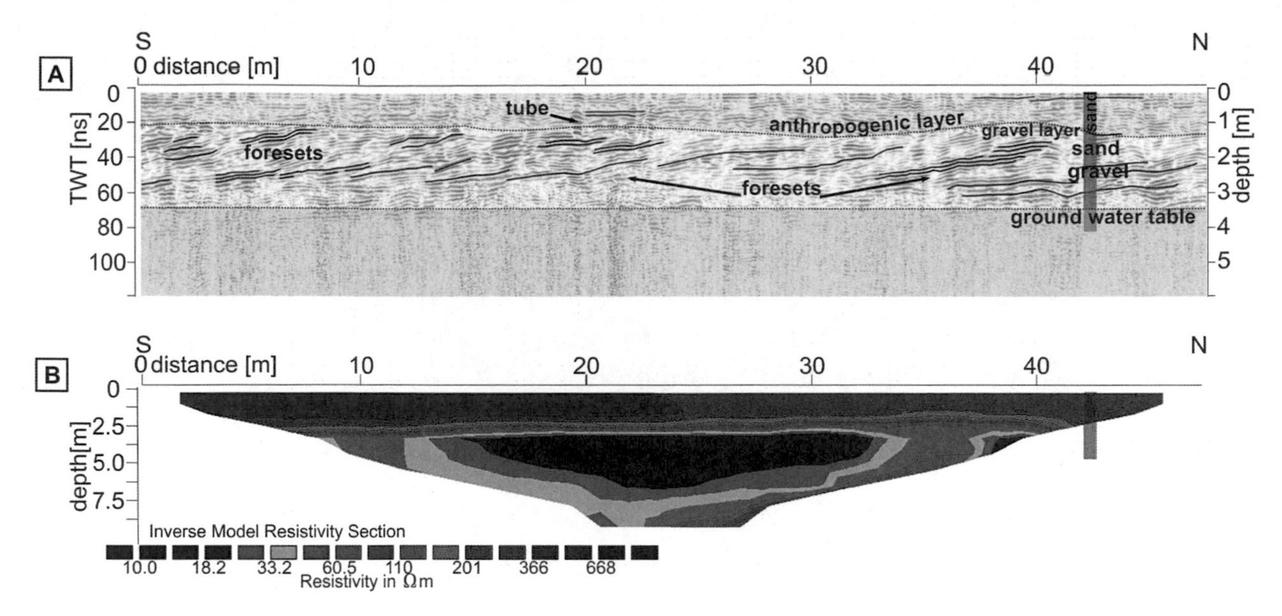

Figure 5.6.: A: Interpreted N-S oriented GPR line no. 79 (270 MHz antenna) at the Struga site with a total length of 47 m. Core U03 and lithologies are projected into the profile. Vertical exaggeration is 1.5. B: Resistivity measurement using Wenner array with an electrode spacing of 1.25 m and a maximum depth of 8.28 m. Three iterations have been applied for inversion with a final error of the model of 2.0 %. Red column shows projected position of core U03. Sudden drop of resistivity values at an approximate depth of 3 m is interpreted as the water table. Modified after Hoffmann et al. (2012).

5.2. Velestovo

5.2.1. Drilling and core logging

The core drilled at the Velestovo site (fig. 5.8) has a total length of 900 cm. The ground water level was encountered at approximately 3.7 m depth.

Four main stratigraphic sequences are apparent:

(1) 900 - 700 cm: Grey-yellowish-whitish marly, partly laminated, open-lacustrine deposits. Littoral faunal and floral remains (e.g. shell fragments of gastropods, tubes of tubifids, ostracod valves and oogonia of Chara sp.; fig. 5.5) are present within the sediments. A high carbonate content (fig. 5.9 indicates a possible influence by carbonate influx of the karstic springs of Biljana (fig. 5.1). A noteworthy sediment section is between 830 cm and 800 cm, which is composed of clayey marls (high proportion of carbonate) with a high content of organic material and gravel (clasts up to 2 cm). This section contains organic remains such as particles of roots and leaves, coaly fragments and seeds, as well as ostracods, oogonia and calcified stems of Chara sp. and was dated to 6520 BP. ± 20.

(2) 700 - 300 cm: Swamp development in a coastal lagoonal environment influenced by occasional fluvial terrigenous input. Several peat sequences are intermitted by silts and sands including fining-upward sequences (e.g. 520 - 502 cm). At 304 cm, a 4 cm thick debris flow layer with a silty matrix containing gastropods and gravel occurs.

Figure 5.7.: Photographs of the recent Chara-belt close to Gradiste (see fig. 5.1). A: Fields of Characeae at a depth of about 8 m. B: Orange coloured oogonia of Characeae in Lake Ohrid.

(3) 300 - 178 cm: Silty and sandy section of terrigenous material containing organic matter and fragmented carbonate shells, possibly originating from a close-by fluvial channel, and representing distal fine-grained deposits.

(4) 178 - 0 cm: Anthropogenically modified horizon with grain sizes between silt and gravel. The entire sequence contains large clasts up to 20 mm in diameter, brick fragments, charcoal and roots. At a depth of 173 - 150 cm, aquatic and terrestrial gastropods were encountered.

The MS signal (fig. 5.8) also reflects the four stratigraphic intervals described above and correlates well with the sequences. However, major positive excursions are observed in the peaty sections. Almost no MS signal was recorded from the marls of the lowest part (700 - 900 cm); only small deviations occur between 830 and 800 cm where values increase. The core material of DA01 is less diverse in its ostracod fauna than core U03 (Candona formosa, C. goricensis, C. lychnitis, Paralimnocythere karamani and P. slavei) and less well preserved, but Characeae remains and fragmented aquatic gastropod shells are also present. XRF data were obtained for the lower 2 m (900 - 700 cm) of the Velestovo core. The elements K, Ca, Ti, Fe, Sr and Zr/Ti have been plotted in figure 5.9. It is obvious that this part of the core is relatively uniform apart from the sequence between 830 cm and 800 cm where the values in all plotted elements vary distinctively. These observations also correlate well with the MS data. The radiocarbon ages of three samples taken at

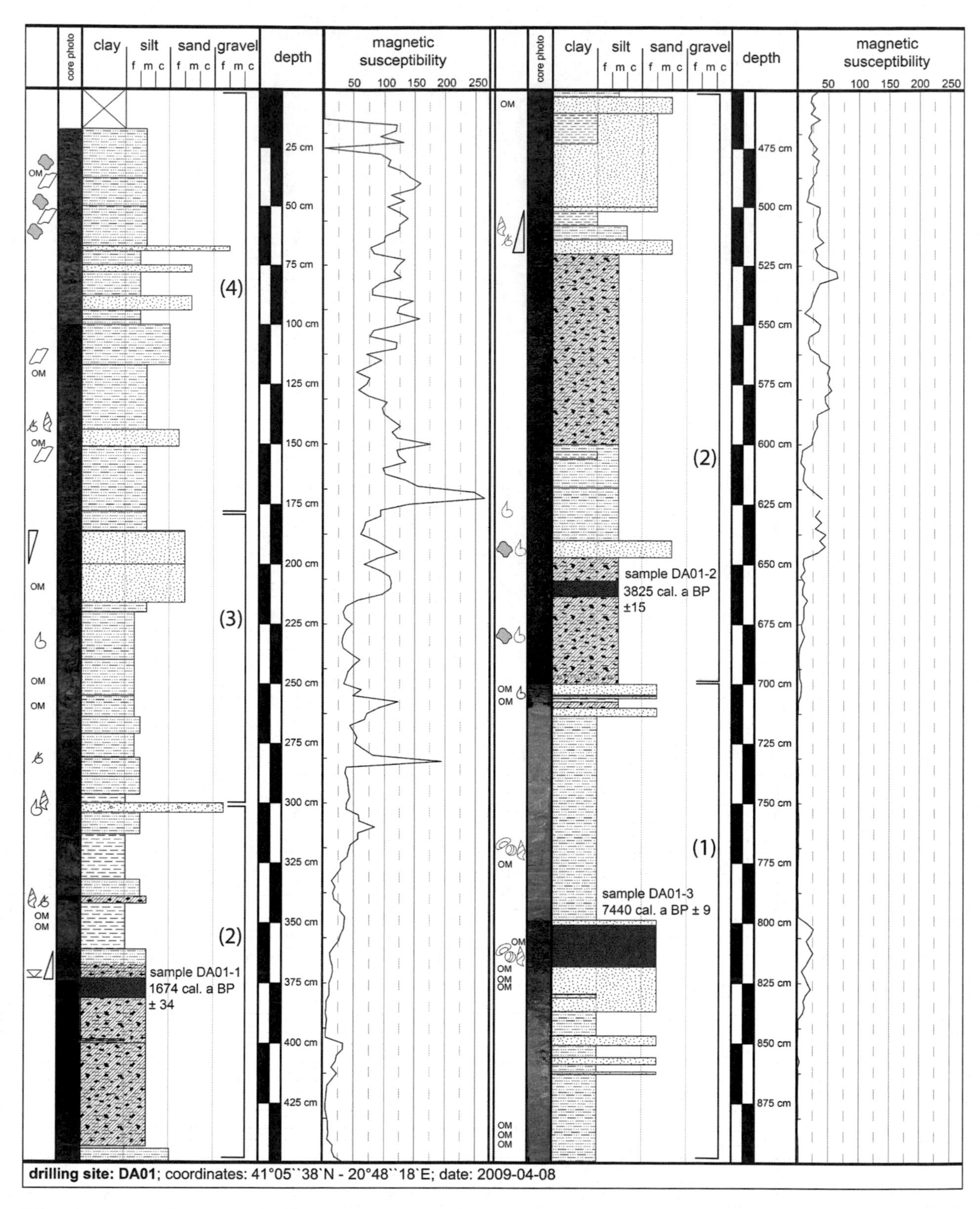

Figure 5.8.: Core log of DA01 with a total length of 9 m. Grain size, fossil content and magnetic susceptibility are displayed. The red signature shows the location and calibrated age of ^{14}C samples. See figure 5.4 for legend. Modified after Hoffmann et al. (2012).

Table 5.1.: Radiocarbon dates of core DA01. Radiocarbon concentrations are given as fractions of the modern standard, $D^{14}C$, and conventional radiocarbon age, following Stuiver and Polach (1977), and calibrated radiocarbon age after Danzeglocke et al. (2013). Sedimentation rates were calculated from the ^{14}C ages assuming that the differing compaction of several lithofacies is neglectible. Modified after Hoffmann et al. (2012).

	DA01-1	DA01-2	DA01-3
depth (cm)	373-381	656-662	800-820
^{14}C conventional (aBP)	1765 ± 15	3540 ± 15	6520 ± 20
^{14}C calibrated (cal. aPB)	1674 ± 34	3825 ± 37	7440 ± 9
sedimentation rates (mm/a)	(0-381 cm) 2,16 ± 0,25	(381 - 662 cm) 1,58 ± 0,18	(662-820 cm) 0,53 ± 0,06

depths of 820 cm, 662 cm and 381 cm are displayed in table 5.1. DA01-3 (7440 cal. a BP ± 9) was sampled in the section 830 - 800 cm which is described in detail above. Samples DA01-2 (3825 cal. a BP 37) and DA01-1 (1674 cal. a BP 34) were taken in peat sequences containing high portions of roots and wood.

5.2.2. Geophysical investigations

GPR profile 05 at the Velestovo site has a total length of 350 m (see fig. 5.2B). Generally weak and mainly parallel reflections are observed, therefore, only 35 m of the entire profile

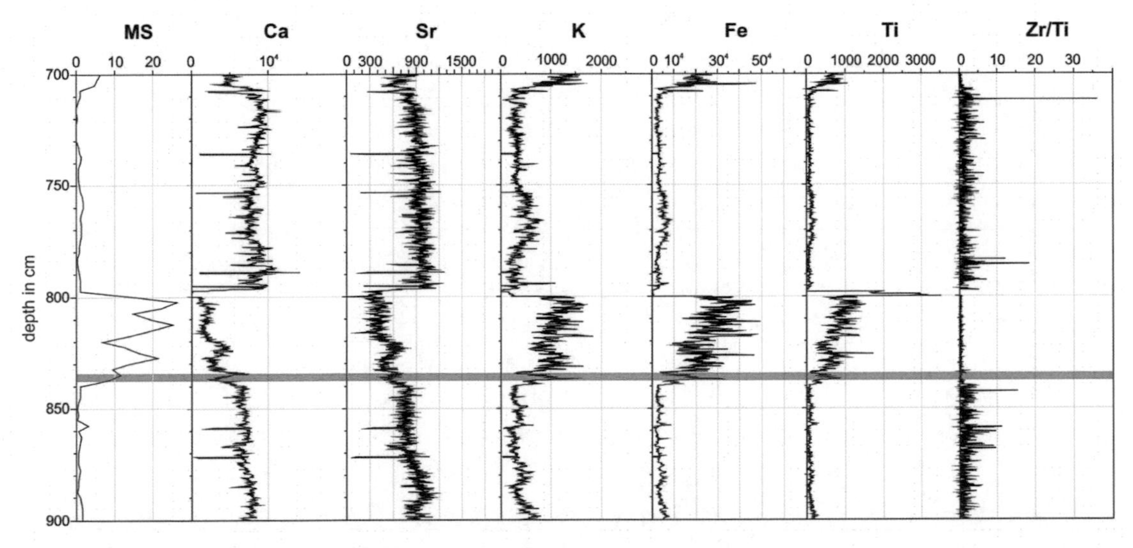

Figure 5.9.: Magnetic susceptibility and XRF-measurements of core DA01 between 900 and 700 cm depth. Element intensities of Ca, Sr, K, Fe, Ti and Zr/Ti are displayed as counts. The red signature refers to the sample at a depth of 836 cm which is described in chapter 5.2. The analysis was carried out at the Institute of Geology and Mineralogy at the University of Cologne using an ITRAX core scanner (COX Ltd.). Scanning resolution was set to 1 mm with an analysis time of 10 s per measurement. The relative concentrations of the elements were derived as an equivalent to the count rates. From Hoffmann et al. (2012).

are shown (fig. 5.10). The topmost 50 cm show parallel reflectors and anthropogenic modifications (e.g. a tube at 260 m and associated road construction). A rugged pattern points to gravel horizons with concave-up reflectors which are interpreted as channel structures (at 237 - 243 m). Below that, a slight inclination of the strata towards the lake (west) is observed (fig. 5.10). At about 75 ns TWT, corresponding to 370 cm depth, the radar signal vanishes; no interpretable data were obtained below this depth which is most likely due to the presence of the ground water table.

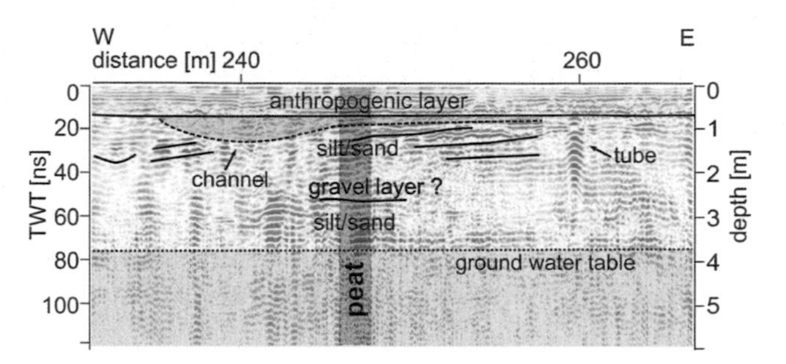

Figure 5.10.: Segment 230 - 265 m in W-E orientation of GPR line 05 at Velestovo (270 MHz antenna; total length 346 m) including interpretation. Red signature shows projected position of core DA01 including lithology. Vertical exaggeration is 2.6. Modified after Hoffmann et al. (2012).

5.2.3. Interpretation

The Velestovo site exhibits a sudden change in sedimentary regime along the Lake Ohrid coast. The formation of the carbonate-rich marly lake sediments of the lowest section (900 - 700 cm can most likely be related to input from the karstic springs of Biljana (Albrecht and Wilke, 2009). As proposed by Lézine et al. (2010), the origin of the calcareous sediments are the limestone outcrops. The carbonates are dissolved by rain water and soil acids and transported into the lake as run-off or by the karstic aquifers. In contrast, the organic and clastic rich interval between 830 cm and 800 cm testifies to a high clastic input at times of high precipitation. This hypothesis is supported by the geochemical data, where highs in the Ti content are interpreted as fluvial clastic input (Vogel et al., 2010a). The lamination observed in core DA01 needs a relatively calm depositional environment which does not presently exist. This is due to the Velestovo site being situated along the east coast of Lake Ohrid which is generally affected by western wind and wave action. However, the geomorphology at this stretch of coast could possibly have allowed the formation of a local lagoon in Holocene times; this would provide such a depositional environment, protected from strong winds and waves. At this site a sedimentation rate of 0.44 mm/a was calculated from the radiocarbon ages. This correlates well with the data of Vogel et al. (2010b) and Leng et al. (2010) who show a sedimentation rate of 0.5 mm/a for the Late Holocene in core Co1202 (located offshore, NW of Velestovo; fig. 5.1). In Lake Ohrid a significant hard water effect on ^{14}C dates was observed by Wagner et al. (2008, 2010) where a sample from surface sediments was ^{14}C dated to 1560 BP. An effect like this in sample DA01-3 would lower the sedimentation rate to approximately 0.3 mm/a at this site. But, as the dated material was only wood, roots and leaves, the hard water effect can probably be neglected. Between the lacustrine marls and the peat deposits, no transitional sediments (e.g. beach sands and gravels) have developed. We interpret the remarkable change from littoral conditions to a coastal swamp as a sudden lake level drop of several meters within the Holocene. This can either be caused by climatic or tectonic influences.

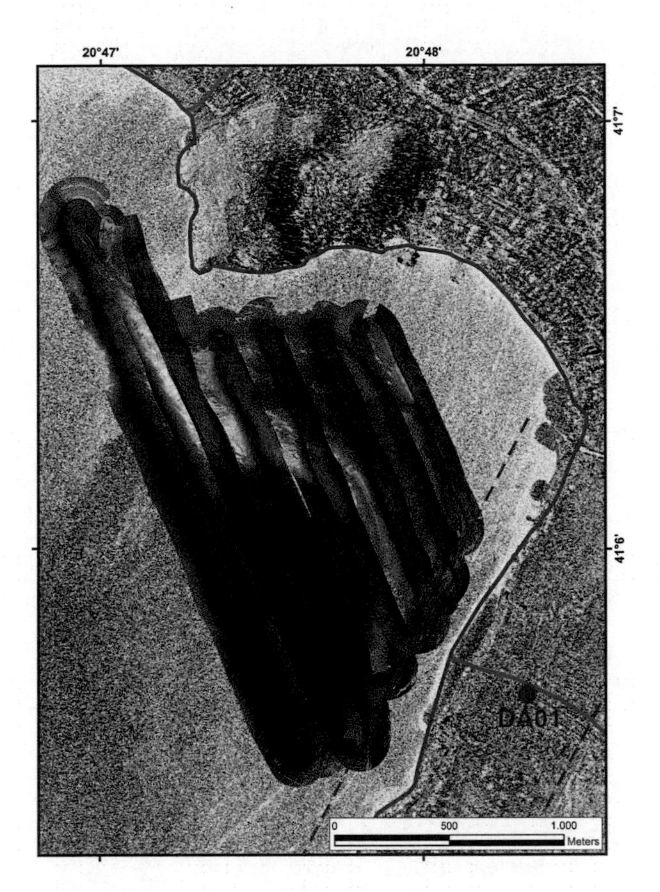

Figure 5.11.: Sidescan sonar data from Ohrid Bay showing fault traces indicated by red dashed lines. The location of core DA01 is projected into the graphic. Light colours indicate high backscatter. For location see figure 5.1. From Hoffmann et al. (2012).

A coast-parallel large normal fault (N30E) stretches off-shore along the wetland plain (see fig. 5.11 and causes a steep decrease in relief within the lake (Wagner et al., 2008). High-resolution seismic data (Lindhorst et al., 2010) suggest very recent activity along that fault. In addition, the Ohrid area is characterised by frequent normal faulting connected to moderate seismicity (Hoffmann et al., 2010; Reicherter et al., 2011). Therefore, the dramatic environmental change is interpreted as fault-induced footwall uplift rather than climatic lake level fluctuations. Lindhorst et al. (2010) also show that the environmental conditions did not undergo a significant change since the mid-Holocene. For the development of this part of the coastline, the following model of lagoon development is proposed. During the movement along a normal fault, which means a downthrow of the hanging wall block, significant coseismic uplift is observed within the footwall of the fault during an earthquake. The L'Aquila earthquake (April 6, 2009) demonstrated that the ratio of the footwall uplift versus the hanging wall subsidence is around 0.3 (e.g. Papanikolaou et al., 2010). Therefore, ongoing faulting, and the development of listric normal faults, led to a tilted block which partly separated a stretch of water from the lake; the separated body of water had a lagoonal like setting. This would create the fitting shallow environment for the deposition of the marls. Later, further development of the basin, and the propagation of faults towards the basin center, created a new fault located further to the west. Due to coseismic hangingwall-uplift, the lagoon has also been lifted up resulting in swamp formation. Later, the levels eroded so that today only parts of the levels are still preserved at the hill of the Inex Gorica Hotel (see fig. 5.2B). The peaty interval refers to periodical wetland formation (lagoon) in the plain south of Ohrid city, probably related to the change of weather conditions and, therefore, a change of sediment input. Below the swamp sediments, sandy-silty deposits of terrigenous origin are interpreted as distal alluvial fan deposits of the Velestovo creek and include periodic flood sediments (debris flows). However, today neither an alluvial fan nor a delta of the little creek is active and the coastline is linear. Besides that, the Velestovo creek is artificially channelised as soon as it enters the plain. Agricultural land, including buried pipes for irrigation, dominate the upper portion of the core.

5.3. Sveti Naum

5.3.1. Drilling and core logging

The recent delta of Cerava River in the south of Lake Ohrid is characterised by a flat delta plain incised by the meandering river including several ox-bows. We drilled three cores in a transect perpendicular to the coastline (fig. 5.2C). Core PL 03, with a total length of 300 cm (fig. 5.12), is described here in detail.

(1) 300 - 200 cm: Sandy sediments containing quartz grains, mica and organic material. The occurrence of white mica is related to the content of organic matter, both of which are enriched with depth.

(2) 200 - 132 cm: Fining-up sequences. Three fining-up sequences are identified (200 - 180 cm, 180 - 170 cm and 160 - 132 cm); all are sandy to silty with clay clasts, organic material and white mica. At 162 cm, a 3.5 cm thick oxidation horizon occurs in between two of the sequences. Towards the base of the whole interval (200 cm depth) the silt fraction decreases, whereas the content of organic material increases.

(3) 132 - 0 cm: Silty to clayey material with no organic remains. In the lower part, the grain size decreases from boulders of quartzite and quartz, with diameters up to 35 mm in the basal layer, to sand/silt. The uppermost 47 cm contain recent soil with roots and a mixed interval of sand and gravel (grain size up to 15 mm in diameter).

PL02 and PL01 open window samples were drilled at approximately 600 m and 900 m (fig. 5.2C) from the shoreline respectively. PL02 was carried out about 7 m above lake level and PL01 was carried out around 10 m above lake level. Both reach a total depth of 600 cm. Lithologies of both are comparable in sedimentary composition to core PL03, with fine to medium-grained sands, silts and occasional gravels.

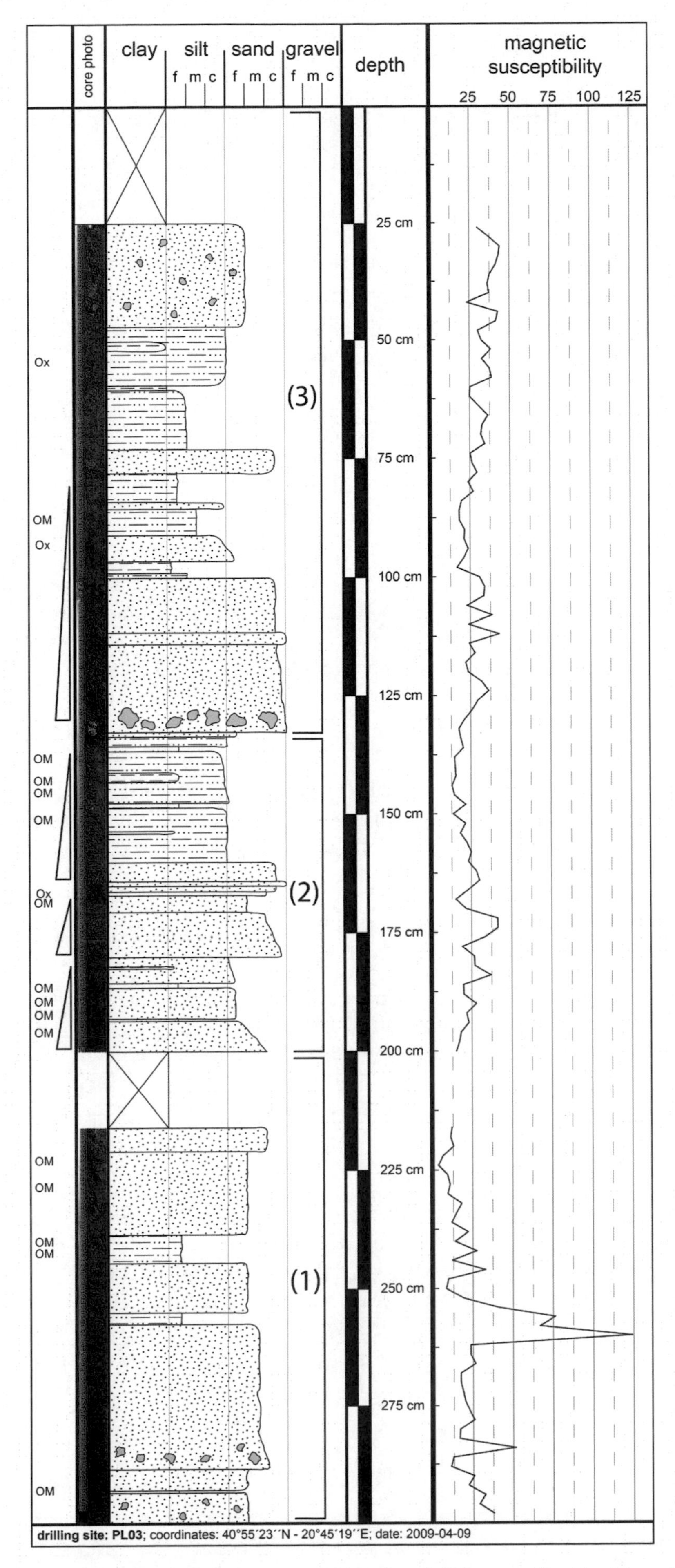

Figure 5.12.: Core log of PL03 with a total length of 300 cm. Grain size and magnetic susceptibility are displayed. See figure 5.4 for legend. See figure 5.4 for legend. Modified after Hoffmann et al. (2012).

5.3.2. Geophysical investigations

A series of 52 GPR profiles and resistivity sections have been carried out at the Cerava delta (fig. 5.2C). Here, two are described as characteristic examples. Profile 56 has a total length of 351 m and is only partly displayed (fig. 5.13). Horizontal reflectors up to a depth of about 100 cm are underlain by northward dipping reflectors (between 50 and 75 m at a depth of 100 - 300 cm) and several concave-up patterns (35 - 40 m and 48 - 52 m at a depth of 100 - 200 cm). These structures can be traced in almost all N - S trending radar profiles (fig. 5.2C) and are interpreted as channels. Profile 32 (fig. 5.14A) shows horizontal reflectors that can be traced over several tens of meters down to about 20 ns. A distinct channel occurs between 180 and 190 m at a depth of about 100 cm. The related electric resistivity profile R1 (fig. 5.14B) shows a significant drop in resistivity values at a depth of approximately 300 cm with a slight dip to the south that correlates with the dipping of reflectors in the radargram. In the very north of the resistivity profile, a concave structure can be observed which correlates well with the concave structure in the radargram.

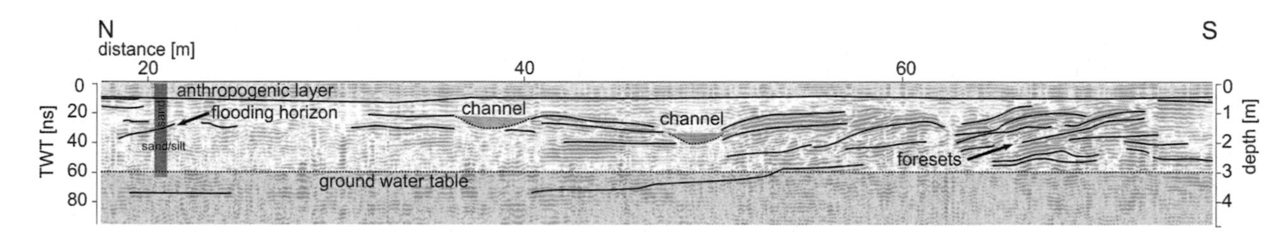

Figure 5.13.: GPR line 56 (270 MHz antenna) at Sv. Naum with a total length of 351 m. Only the segment 18 - 79 m in N - S orientation is displayed and includes interpretation. Red column shows approximate position of core PL03 (fig. 5.12) with projected lithologies. Vertical exaggeration is 1.5. From Hoffmann et al. (2012).

5.3.3. Interpretation

Drillings at the Cerava delta provided evidence that the delta is dominated by alluvial plain deposits. These are characterised by coarse to fine-grained clastics with occasional soil development (palaeosols). Northward dipping reflectors are interpreted as shallow deltaic foresets. Abundant mica and granitic pebbles testify to a provenance of either basement rock which outcrops to the south along the Greek/Albanian border, or more likely of conglomerates of the Mirdita Zone (see Hoffmann et al., 2010) which are exposed to the south and west of Pogradec (see fig. 5.1). The correlation between the occurrence of mica and organic material is interpreted as the influence of higher precipitation on the transportation capacity of the Cerava River. The Cerava delta plain is used for agriculture and has been incised by the meandering Cerava River, leaving several channels of different ages. Because no remains of Chara sp. or ostracods have been encountered in the core, there is no evidence for lacustrine sediments in that part of the lake during the Holocene.

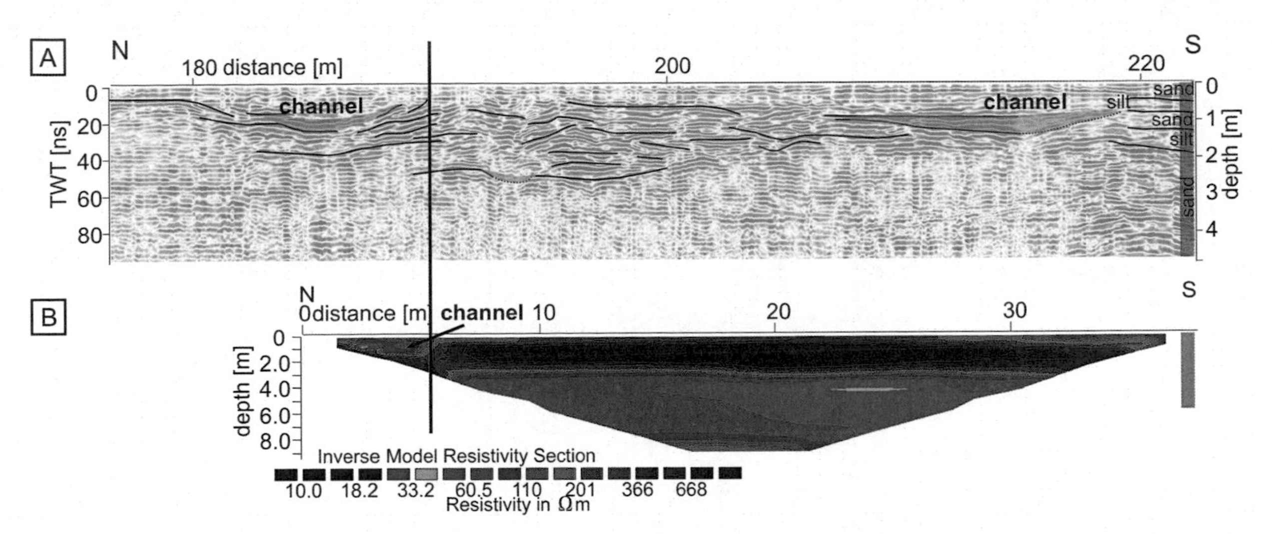

Figure 5.14.: A: Interpretation of segment 178 - 222 m of GPR line 32 (270 MHz antenna; total length 224 m). Red column shows position of core PL02 (see fig. 5.2C) with projected lithology. Orientation of the profile is N-S. Vertical exaggeration is 1.5. Black line marks the beginning of a channel structure in both profiles. B: Resistivity line R1 in Dipole-Dipole array with an electrode spacing of 1.00 m and a maximum depth of 9 m. Red column shows projected drilling site PL02. Three iterations have been applied for inversion with a final error of the model of 5.50 %. From Hoffmann et al. (2012).

5.4. Daljan River Delta

5.4.1. Drilling and core logging

At the Daljan River delta two locations have been drilled, each with the open window sampler and liner cores (see fig. 5.2). At DL01, a liner core with a length of 3 m (DL01) and a 5 m open window sampler (DL I) were recovered. At DL02 the open window sampler was drilled up to a depth of 5 m (DL II), whereas liner core DL02 (fig. 5.15) reached 4 m depth. Here, core DL02 drilled close to the shoreline at the Daljan delta, is described in detail (fig. 5.15). The content is relatively uniform and shows mixed clastic sediments that indicate a deltaic environment. The core can mainly be subdivided into an anthropogenically influenced upper part and underlying clastic sequences which contain mainly fine gravels, sands and silts.

(1) 400 - 54 cm: At the bottom of the core (393.5 - 395 cm) layers of intercalated sand and silt occur. Above this the core is composed of fining up sequences at 361 - 384 cm, 163 - 137 cm, and 100 - 70 cm depth (see also fig. 5.16). These are made up of gravel, sand and silt with well sorted and subangular to subrounded grains. A thick coarsening upward section is detected between 300 - 186 cm.

(2) Anthropogenic remains such as brick fragments characterise the upper part of the core (54 - 0 cm).

The magnetic susceptibility reflects the anthropogenic influence in the upper 54 cm with slightly higher values. Other highs occur in unsorted coarse grained sands and gravels.

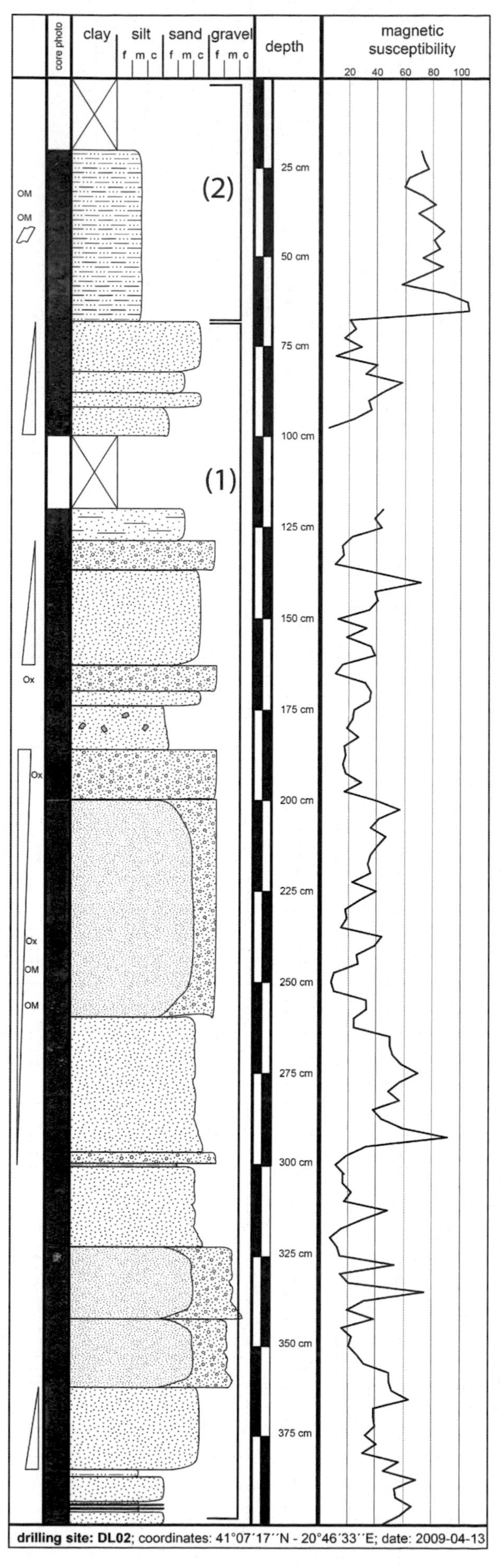

Figure 5.15.: Core log of DL01 with a total length of 400 cm. Grain size and magnetic susceptibility are displayed. See figure 5.4 for legend.

Overall, the signal is relatively uniform with values between 40 and 60 SI.

The liner core DL01 distributes a comparable lithology to DL02 with lake facies in the lowest 12 cm and fining up and mixed clastic sequences above, representing fluvial facies. Red/brown oxidation horizons (86 cm, 142 cm, 146 cm, and 172 - 200 cm) are evidence of ground water influence, possibly by seasonally alternating lake levels. The open window sample DL II is the only core with laminated clay/silt layers at the bottom which are interpreted as shallow lake sediments. Fossils were not found in any of the cores.

Figure 5.16.: Fining upward sequence (100 cm - 80 cm) in core DL01. Photograph by Katharina Wohlfahrt.

5.4.2. Geophysical investigations

At the Daljan delta four GPR lines were carried out; no resistivity measurements were undertaken at this location. The lines horizontal to the shoreline do not show any significant structures except a few concave-up patterns interpreted as channels. In contrast, radar lines 01 and 03 show some interesting features. Here, the southernmost 100 m of line 01 is displayed (see fig. 5.17). Southward dipping reflectors are apparent which become more distinct towards the lake shore; above and below, horizontal layers can be depicted. According to the GPR profiles, the ground water table is located at a depth between 4 and 5 m, where the radar signal diminishes.

5.4.3. Interpretation

The data from the Daljan delta with mixed clastic sequences of mainly sand and gravel represents fluvially dominated delta type facies and sedimentary architecture. Prograding delta sequences can not only be seen in the coarsening upward sections of the cores, but also in the foreset structures pictured by the southward dipping reflectors of the GPR profiles. The poorly sorted material of the coarsening up layers is an indicator for an unsteady depositional environment as it occurred in a river dominated delta. Well sorted (concerning grain sizes) fining upward sequences are interpreted as overbank sediments or cut-off meanders. The coast parallel GPR profiles both reflect asymmetric channel structures that are typical for meandering rivers. Above these, the horizontal reflectors point to ongoing sedimentation under terrestrial conditions with soil development in the upper horizon. Therefore, the delta is considered to be still actively prograding. No evidence for lake transgression was found in either the cores or GPR data.

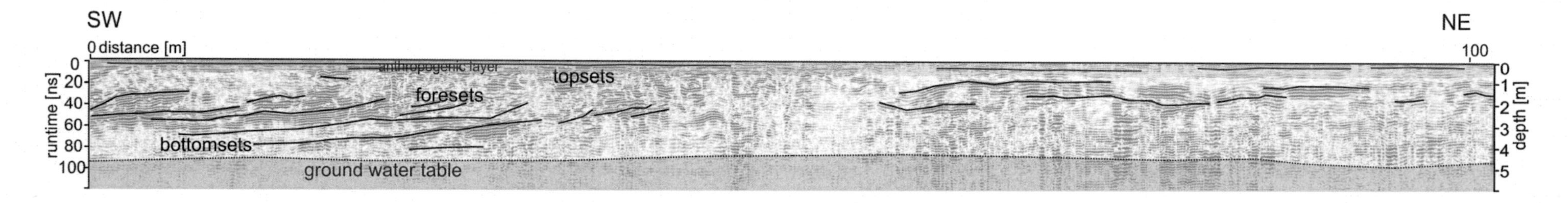

Figure 5.17.: Interpreted GPR line 1 (270 MHz antenna) from the Daljan River delta with a total length of 103 m. Vertical exaggeration is 1.5.

5.5. Lini

5.5.1. Drilling and core logging

Two cores have been recovered at the Lini Peninsula, core L01 with 2 m length close to the lake shore and core L02 with 9 m length at the center of the peninsula in agricultural land (see fig. 5.2).

Core L01 (see fig. 5.18) was subdivided into three sections with specific sedimentological characteristics.

(1) 200 - 124 cm: Mainly unsorted, grain supported gravel layers in sandy matrix. Grains are mostly angular limestone fragments from about 2 cm to 4 cm in size, most probably originating from the nearby limestone cliffs (see fig 5.2).

(2) 124 - 70 cm: A sandy section with organic material and fewer gravel grains makes up the lower part (124 - 100 cm); within this a clay layer occurs at 104 -105 cm. Above, a sand layer has developed characterised by light grey carbonate debris, a shell sample at 85 cm and rounded pebbles of 2 cm in size. This is interpreted as beach development supported by the peat sequence formed at 82 - 75 cm, marking the transition from lake to land. An organic rich layer with strongly dissolved shells lies on top.

(3) The uppermost 70 cm are composed of brown soil. The lower 25 cm of this section is organic rich sandy/silty soil with light brown reduction spots. The water table lies at 55 cm on top of this reduction horizon. The upper part of the core is composed of compacted recent soil (36 cm - 55 cm).

Core L01 shows a coastal silting up process with mixed clastics. Mainly angular limestone gravels in sandy matrix (a short transport distance is reflected in unrounded and unsorted material), originating from the Lini limestone block, are located at the bottom of the core. A tidal zone with beach development and swamp formation is present in the central part (indicated by rounded pebbles, organic material and broken shells) and finally soil formation has developed in the upper part.

The content of core L02 (see fig. 5.19) is relatively uniform. The clayey sections are composed of carbonate-poor loess loam. The ground water table was discovered at 414 cm. Four stratigraphic intervals were designated.

(1) 900 - 662 cm: Ochre coloured clay section.n Only a few exceptions from the ochre horizons can be seen (e.g. 858 - 869 cm and 772 - 800 cm) where the colour changes to dark grey. Lime and manganese concretions, as well as ophiolite fragments, can be found at various depths.

(2) 662 - 351 cm: Mixed sand and silt sediments containing lime and ochre concretions at different depths. At 472 - 475 cm a manganese horizon has accumulated. Above, sandy layers containing few well rounded gravels are separated by a clay section.

(3) 351 - 58 cm: Loam dominated material with manganese, lime and ochre concretions of up to 1 cm in diameter at various depths.

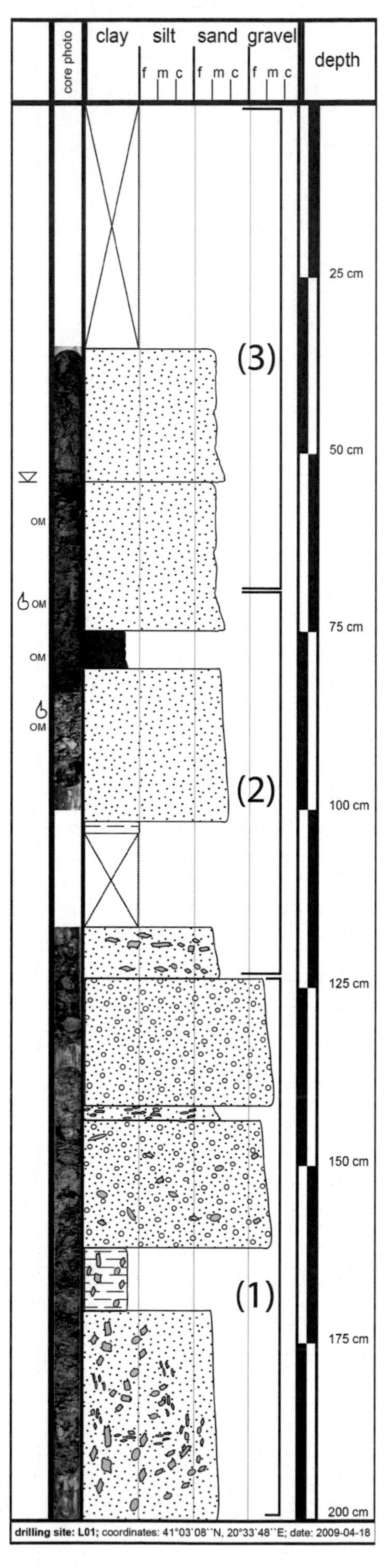

Figure 5.18.: Core log of L01 at Lini with a total length of 3 m. Grain size and special features are displayed.

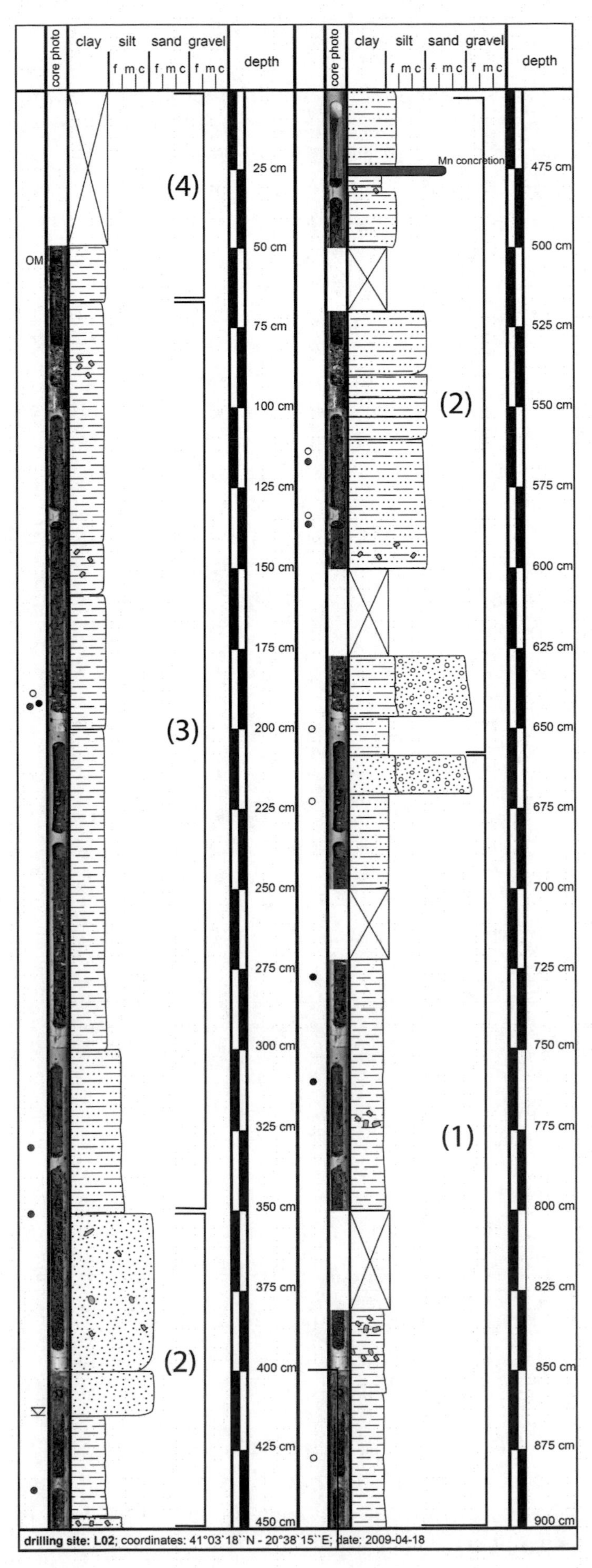

Figure 5.19.: Core log of L02 at Lini with a total length of 9 m. Grain size and special features are displayed.

(4) The upper 58 cm are composed of loamy soil with roots.

Core L02 is a profile of a hydromorphic carbonate-poor loess soil characterised by Mn, Fe, and Ca concretions at various depths. This is typical for pseudogleys which are subject to periodically stagnating surface water. Rain water cannot infiltrate into the compact clay; therefore, rain leads to a high water content and a perched water table. The water in the pore space displaces aerial oxygen, and, together with aerobic bacteria, reduction conditions prevail. The oxidated Fe and Mn ions show a better solubility and diffuse into adjacent areas where they get into contact with air filled pore spaces. This leads to precipitation and accumulation of iron and manganese oxides and hydroxides leading to small concretions spread over large areas and depths. The reduction horizons are, therefore, bleached. This can be seen in lighter horizons which are not as red (100 - 149 cm, 210 cm, 400 - 414 cm, 520 - 540 cm, 627 - 646 cm) in this core. Mottled horizons, like the one from 475 to 500 cm, are typical mixed reduction/oxidation zones. Although the soil profile is not very typical for a gleysation process, a distinct number of indications were found which characterise this soil as pseudogley or stagnosol. These include the flat morphology of the Lini Peninsula and the loess soil, which can be, together with excessive water input, a feedstock for this type of hydromorphic soil. The lower part of the core (below 772 cm) is made up of dark grey sediments, which points to a reductional milieu as it is deposited in a lake environment.

5.5.2. Geophysical investigations

A total of 32 GPR profiles were done at Lini Peninsula with the 270 MHz antenna. Most of them are of poor quality or show mainly horizontal reflectors. Therefore, only two profiles (no. 23 and no. 13) are presented here.

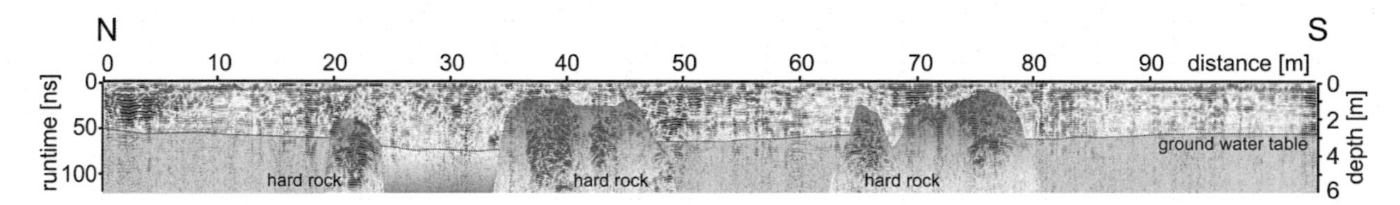

Figure 5.20.: Interpreted GPR line 23 (270 MHz antenna) at Lini with a total length of 102 m. Vertical exaggeration is 1.5.

In contrast to other radar lines, GPR line 23 (see fig. 5.20) shows not only horizontal layers, but also unsteady reflectors between them which are deeper than 1 m. The reflector depth, horizontal extent and hyperbolic pattern leads to the conclusion that these signals portray blocks of limestone bedrock. The adjacent limestone cliffs of Lini is most likely the source of these blocks. It is unclear whether they are solitary blocks, detached from the cliff by rockfall, in-situ limestone bedrock that tapered, or horst and graben like structures that are subject to extension. Further investigation would be necessary to determine their origin. The ground water table is assigned to 2 - 3 m depth according to the extinguishing signal. Some zones (e.g. 50 - 60 m) also have very weak signals between or above well defined reflectors. The cause can be the generally high water saturation and/or water saturated horizons (such as a perched water table) in the loamy soils.

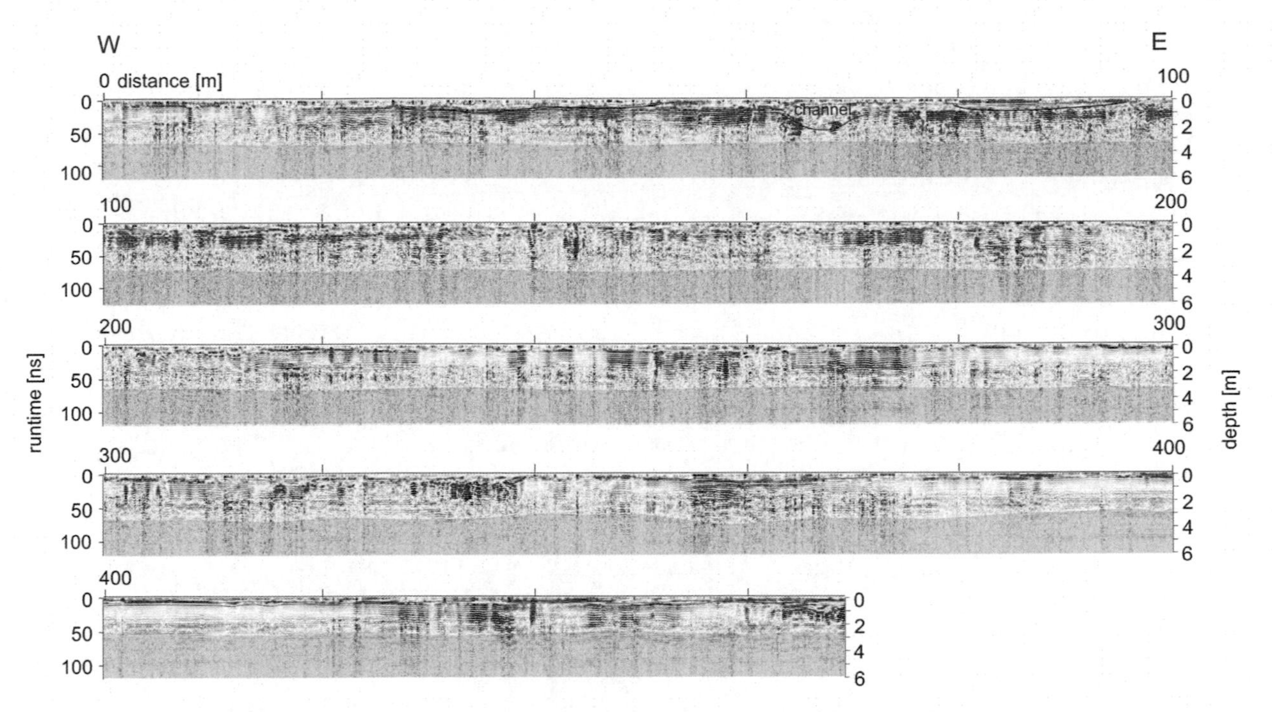

Figure 5.21.: GPR line 13 (270 MHz antenna) at Lini with a total length of 464 m. Vertical exaggeration is 1.5.

GPR line 13 (see fig. 5.21) has a total length of 464 m. Mainly horizontal deposition can be seen in the profile. Exceptions include a channel structure at 70 m and two shallow concave-up structures in the uppermost layers at a distance of about 40 - 60 m and 85 - 95 m. These structures can develop by erosion as a consequence of heavy rain or flooding (high lake level, storm surges or overfull drainage trenches) which cause water build up and run-off on top of the water saturated soils. The signal becomes very weak at a depth around 2 m which is normally interpreted as the ground water table. However in this case, the ground water table has been located much deeper in the drill cores; therefore, this could be the beginning of a zone of saturated soils or a perched water table which can cause the radar signal to fade in the first few meters, although the actual ground water level is not yet reached. The anthropogenic horizon cannot be determined from profile 13. Firstly, this is due to the very homogeneous material throughout the soil column, and secondly, as the soilwork is still traditionally done by ploughing with oxen, only the uppermost 30 cm of topsoil is worked. Therefore, there is not much anthropogenically influenced disturbance to the ground.

Resistivity measurements at Lini Peninsula show two totally different profiles. The L1 profile (see fig. 5.22A) is characterised by generally low values that increase slightly to the west. These are most probably caused by a high ground water table and water saturated soils. The L3 profile (see fig. 5.22B), in contrast, shows a continuous increase of resistivity with depth mirroring undisturbed sedimentation in the center of Lini Peninsula as seen in GPR measurements.

5.5.3. Interpretation

The Lini Peninsula is a halfgraben with horst and graben structures in the subsurface bedrock which is bordered by limestone and ophiolitic hard rock blocks to the east and west. In between these, a flat plain of aeolian silt accumulated which later formed a pseudogley soil under the influence of extensive water input. The moist ground is, therefore, drained by drainage ditches. The traditional ploughing techniques lead to the formation of arched fields which support the drainage of the topsoil. The areas close to the limestone cliffs suffer from the limestone fragments detaching themselves from the cliff and leading to aggradation. GPR and ERM also show mainly parallel sedimentary architecture with some irregularities due to the drainage system and small channels.

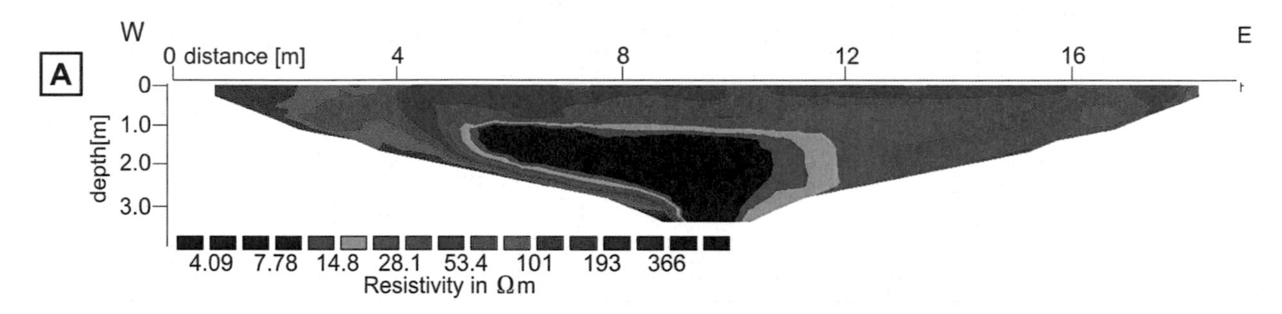

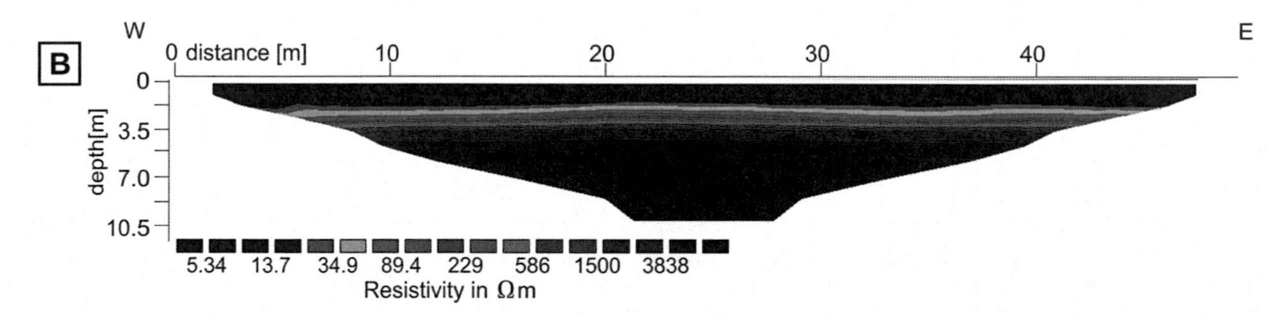

Figure 5.22.: Resistivity measurements of Lini Peninsula. A: L1 Resistivity measurement in Wenner array with an electrode spacing of 0.5 m and a maximum depth of 3.31 m. Three iterations have been applied for inversion with a final error of the model of 3.9 % B: L3 Resistivity measurement in Wenner array with an electrode spacing of 1.3 m and a maximum depth of 9.60 m. Three iterations have been applied for inversion with a final error of the model of 17.6 %.

5.6. Discussion

Shallow drillings and geophysical investigations at five different locations at Lake Ohrid provide an insight into the Holocene coastal evolution of the lake. In general, two main geomorphological systems with different mechanisms driving the sedimentation process can be distinguished. These are: (1) the plains to the north and south of Lake Ohrid, and (2) the steep hillslopes to the east and west belonging to the Mountains of Galicica, Mali I Thate and Mokra. The northern part around Struga is characterised by an extensive shallow plain with large, but very shallow, alluvial fans (fig. 5.1). Their source areas are the calcareous mountain ranges located to the west and east of the plain. Today, these alluvial fans are mainly inactive. At a depth of two meters, there is clear evidence of diverse lacustrine fauna and flora, represented by shell fragments of gastropods, ostracod valves and oogonia of Chara sp., testifying to lake deposits. Findings of oogonia of charophytes are not suitable to infer the exact water depth from sediment archives, as they are dispersal agents and the charophyte communities are highly dynamic (Wagner et al., 2010). Thus, the so-called Chara-belt (after Albrecht and Wilke, 2009) in modern Lake Ohrid at 3 - 20 m water depth (see also fig. 5.7) does not necessarily serve as a modern analogy for past lake level conditions. Shallow, lake-ward dipping foresets in the lower part and typical fluvial facies in the upper part infer a southward sediment transport by a former inflow. The Sateska River is today channelised (since 1962) and flows partly into the Crni Drim River and partly into Lake Ohrid, where it acts as the only northern inflow. According to Jordanoski et al. (2010), the original inflow of the Sateska River used to be east of Mislesevo (fig. 5.2A).

Since the 18th century, the area around Lake Ohrid was successively deforested. The increasing sediment load of the Sateska River led, over time, to a backfilling of the river bed and to a change of the river course. Remnants of the old river bed and land use patterns are indications for the former location of the Sateska River, but are today superimposed by agricultural use and new settlements. In general, the plain is dominated by clastic input; however, this varies over time influenced by climatic variations (see Wagner et al., 2010). Most probably the plain was once part of the deeper basin and has been separated by displacement along faults in NW-SE or WNW-ESE orientation (for more details on basin development see chapter 4).

The Velestovo site, in contrast, exhibited a coastal marsh/lagoon environment during the Holocene, weakly influenced by clastic input of the Velestovo creek. The clastic interbedded sequences are interpreted as periodical flood sediments. The most striking observation at the site is that marly lake sediments are overlain by a thick sequence of peat, representing a sudden change in the depositional environment. Causes for this may be a rapid lake level drop or neotectonic activity of a nearby active normal fault, which runs parallel to the coast (see fig. 5.11). However, where the ostracod remains are present (U03 and DA01), they clearly indicate a lacustrine assemblage in the shallow littoral zone of the lake system (Mikulić and Pljakić, 1970). The Characeae and gastropod fragments are supporting this scenario.

The Cerava plain and the Daljan River Delta constitute typical delta plain depositional environments in the northern and the southern part of Lake Ohrid, including meandering fluvial systems with incised channels. Albrecht and Wilke (2009) describe several prograding delta deposits detected in offshore multichannel seismic data close to Sveti Naum. We can, therefore, conclude that this inflow was stable at least during Holocene.

For the Daljan River Delta no evidence was found for a retrogradation of the delta. Several meanders at different positions were detected which is another indication for a firmly located delta, fed by a river where meanders are relocated over time.

The Lini Peninsula is limited by faults forming a graben structure in which aeolian sediments accumulated. The reasons for seasonal flooding are various and are not only caused by lake level shifts or storm floods.

Our investigations suggest that the northern and southern shores of Lake Ohrid are dominated by sedimentary input. Alluvial fans and deltas are interpreted to be climatically driven, whereas the eastern and western shores (Velestovo and Lini Peninsula, fig. 5.1) clearly show the influence of active tectonics on the depositional system. On the other hand, the annual periodic lake level fluctuations of Lake Ohrid do not exceed 1 m (amplitude; i.e. $\pm$ 50 cm, (Matzinger et al., 2006a; Popovska and Bonacci, 2007). Several high stands with coastal flooding have been reported during the last century (e.g. in 1936 and 1956 Stojardinovic, 1969). Those short-term events flood the coastal areas and form intermittent lagoons (fig. 5.1) which do not reach 695 m a.s.l. In figure 5.1 flooding areas with an increasing lake level were modeled. Yellow coloured areas are at the level of the modern water level and therefore are subject to a high water table and swamp formation. These areas are drained by a vast network of drainage trenches to develop the land as acreage. A raising water level of only 5 m would cause the flooding of wide areas of the plains. The Struga Plain and the Pogradec Plain have been filled with sediments during the Holocene.

The extent of the palaeolake and, therefore, the initiation of lake building is still unclear. These questions could not be addressed by using the applied methods. Deeper drilling and/or extensive deep electric resistivity measurements might produce better results and could help to understand how the lake was formed.

In conclusion, based on sedimentological and palaeontological investigations supported by shallow geophysics, it was possible to distinguish between the northern and southern plains of Lake Ohrid which are dominated by fluvial clastic input, and the eastern and western shores which are controlled by active normal faulting. The rapid changes from open lacustrine to swampy areas are attributed to syntectonic footwall uplift, possibly induced by a palaeoseismic event.

6. Geomorphology

Large earthquakes leave their imprint in the landcape. As the slipped area of the fault plane often extends to the earth's surface. The threshold for ground rupture and fault scarp development are earthquakes with M 5.5 or greater; (McCalpin, 2009) so that over an extended period of time, recurrent large morphogenic earthquakes create a "seismic landscape" (Dramis and Blumetti, 2005; Michetti et al., 1995; Michetti and Hancock, 1997; Michetti et al., 2005; Serva, 1995) or a "palaeoseismic landscape" (McCalpin, 2009). Palaeoseismologists and tectonic geomorphologists use these often short living features to identify active faults and to parameterise palaeoearthquakes (slip rates, subsidence and erosion). Seismic-hazard assessment and seismic-risk mitigation efforts demand a scientific understanding of both the primary (surface rupture) and secondary (surface shaking) effects of the earthquake process, and also the structures and landscapes created by ancient earthquakes (see Michetti et al., 2007).

The Lake Ohrid (fig. 6.1) area hosts a wealth of morphological expressions associated with the seismic activity of the region. These most common features are linear bedrock fault scarps which occur on land as well as in the lake. They give the rapidly rising relief on both sides of the lake a staircase-like appearance (fig. 6.2A). Other well preserved morphotectonic features are wind gaps (fig. 6.2B), wine-glass shaped valleys and triangular facets (fig. 6.2A).

Bedrock or hard rock fault scarps are long-lived expressions of repeated surface faulting in tectonically active regions. They are quite common in semiarid areas, where erosion and sedimentation cannot outpace the fault slip. Post-glacial (or Late Pleistocene, after the Late Glacial Maximum) hard rock scarps at active basin margins are quite frequent in the Mediterranean region (e.g., mainland Greece: Armijo et al. (1991, 1992); Roberts (1996); Roberts and Ganas (2000); Stewart (1993); Crete: Caputo et al. (2006); Italy: Papanikolaou and Papanikolaou (2007); Papanikolaou et al. (2005); southern Spain: Reicherter et al. (2003). They are usually very easy to recognise because they offset mountain slopes, which have been eroded by intense Late Pleistocene weathering and cryogenetic processes (Caputo et al., 2006). Preservation of several meter-high coseismic fault scarps is a function of reduced production and mobility of sediments along the slope, persistent climatic conditions, and cumulative earthquake events along the same fault (tectonic slip rate > erosion rate; fig. 6.4). During glacial conditions enhanced sediment mobility was faster than fault slip movement; no or only minor scarps developed (Papanikolaou et al., 2005). This is broadly confirmed by cosmogenic dating of fault scarps in Greece (Sparta Fault, Peloponnesus, Benedetti et al. (2002); Kaparelli Fault, Greece, Benedetti et al. (2002) and Italy (Magnola Fault, Apennines, Palumbo et al. (2004), where oldest exposure ages of around 20-13 ka were determined for exhumed limestone fault scarps. The well preserved hard rock fault scarps in the Lake Ohrid Basin, especially those made

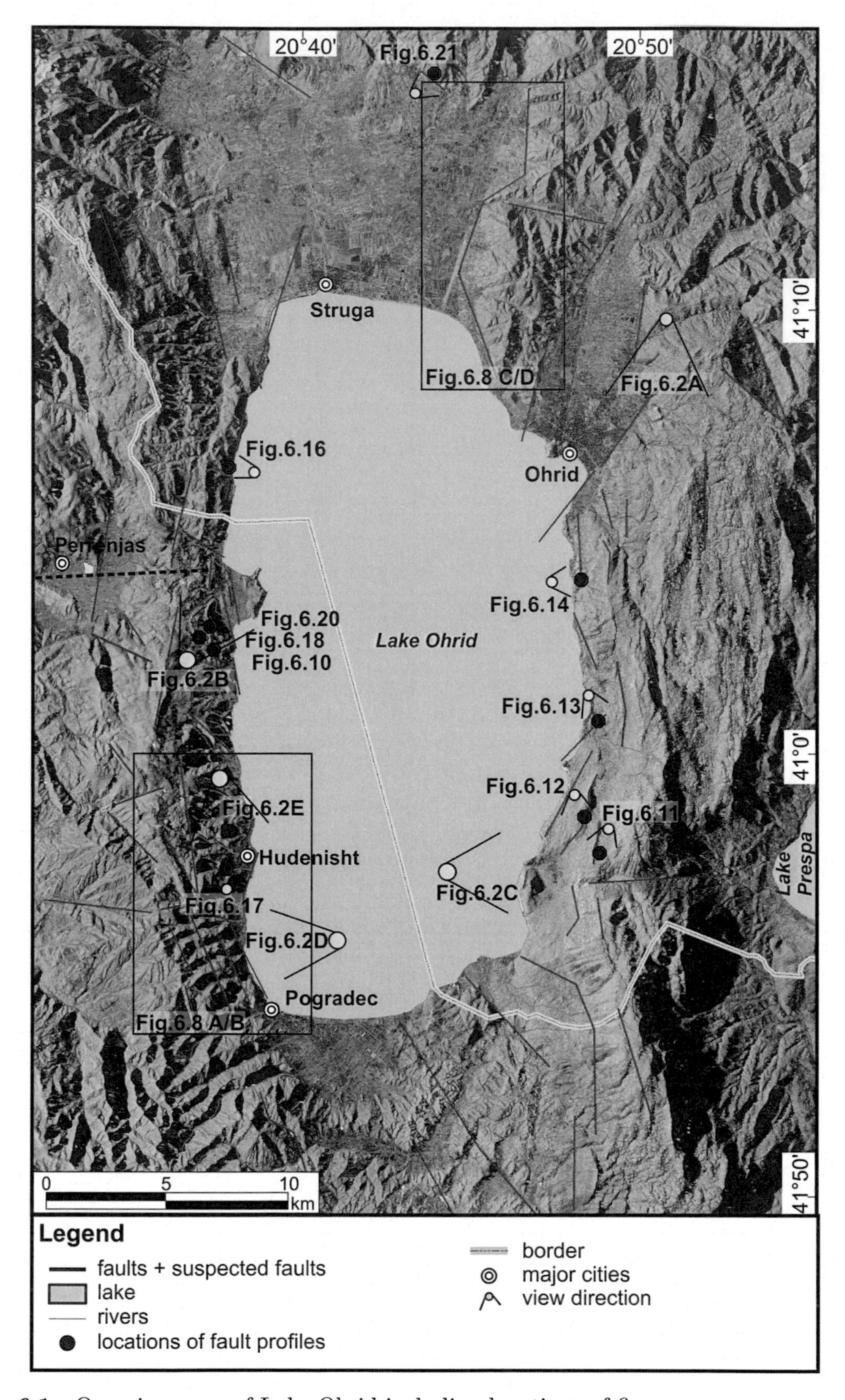

Figure 6.1.: Overview map of Lake Ohrid including locations of figures.

Figure 6.2.: Field photographs of geomorphological features. For locations see figure 6.1. A: Staircased landscape at the slope of Galicica Mountains. Arrows show orientation of normal faults. View to south. B: Panoramic view of Lini Peninsula and Mokra Mountains showing tectonic features. **O**= ophiolites, **C**=carbonates, arrows indicate orientation of normal faults. View to southeast. C: E-W trending wind gap at Galicica Mountains, probably developed as an old fault or discharge connection between Lake Prespa and Ohrid. Black line shows fault trace. View to east. D: Triangular facets on the west coast (Hudenisht area). View to west. E: Panoramic view of the south-western coast. Dashed line shows inferred fault traces (see also fig. 6.1). Note the fault-cut alluvial fan. View to south.

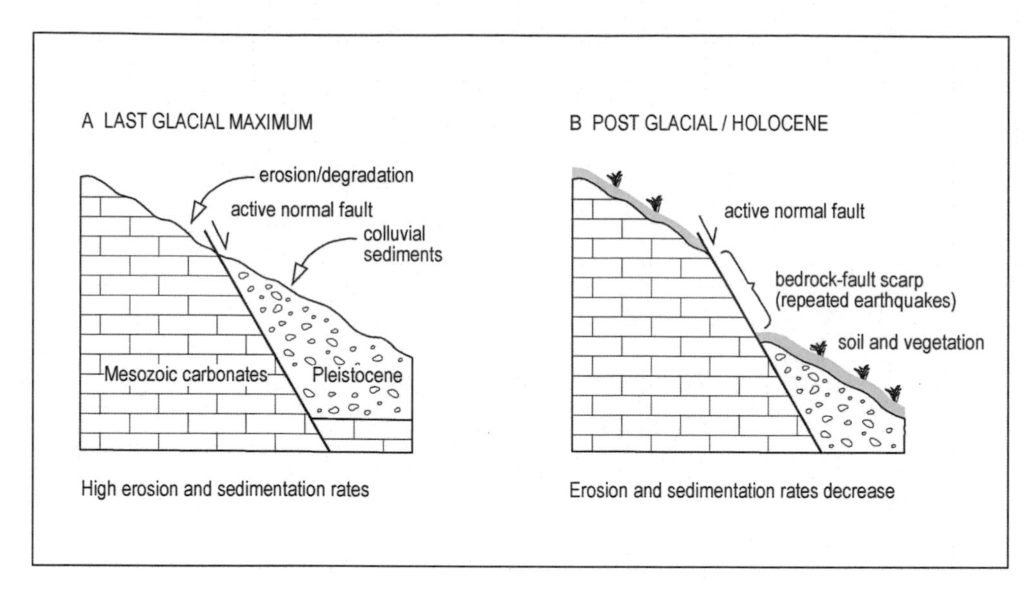

Figure 6.3.: Schematic postglacial evolution of bedrock fault scarps. From Reicherter et al. (2011).

up of limestones and displaced against Quaternary sediments, are also considered to be post-glacial (or Late Pleistocene) after Papanikolaou et al. (2005). Depending on the lithology, short-living topographic features in non-consolidated sediments develop, which are immediately subject to erosion. On the other hand, long-lasting fault scarps in the Ohrid area result in differential erosion on the foot- and hanging walls of faults. These linear features often form coseismically during the earthquake, leaving ribbons of hard rock scarps that form the borders of a basin.

Scarp morphology and preservation is also dependent on uplift and erosion or subsidence and deposition (McCalpin, 2009). This affects both hardrock fault scarps and those in unconsolidated sediments. Where the displacement rates exceed the rates of these geomorphic processes, palaeoseismic landscapes form (fig. 6.4). Obviously this process is highly dependent on the climatic conditions, i.e. humid vs. arid conditions or glacial vs. interglacial conditions. Scarps can be eroded and destroyed or buried by sediments (fig. 6.4-1). Only relicts of the fault scarps remain preserved either in the footwall (fig. 6.4-2) or in sediments of the hanging wall (fig. 6.4-3). A given constant displacement rate of 0.1 mm/a (including an error, see figure 6.4 black dot) on a normal fault will lead to the destruction of a fault scarp (sector 1) with erosion rates larger than 0.1 mm/a (uplift). If erosion rates are smaller than 0.1 mm/a, partial preservation of a topographic step due to a normal fault (sector 2) is expected. If deposition or subsidence occurs, the hanging wall of the fault contains partially preserved sediments (sector 3). In sector 4 the fault scarp is buried by enhanced sedimentation. As an example, blue and red dots show effects on the scarp preservation if sedimentation rates are accelerated, and are larger than the fault displacement rate. Thus, for the scarps at Lake Ohrid Basin, the relationship between displacement rate vs. deposition in sector 3 is assumed, whereas the scarps of postglacial stage fall into sector 2. However, it is evident that hard rock scarps are more resistant to erosion than scarps in unconsolidated sediments, and higher depositional rates occur within the hanging wall area. This contrasts with the scarp development model of McCalpin (2009) and includes a relative age of the scarps (e.g. fig. 6.5) and stepped scarps. Taking this into account, generally younger scarps are to be observed, i.e. younger active

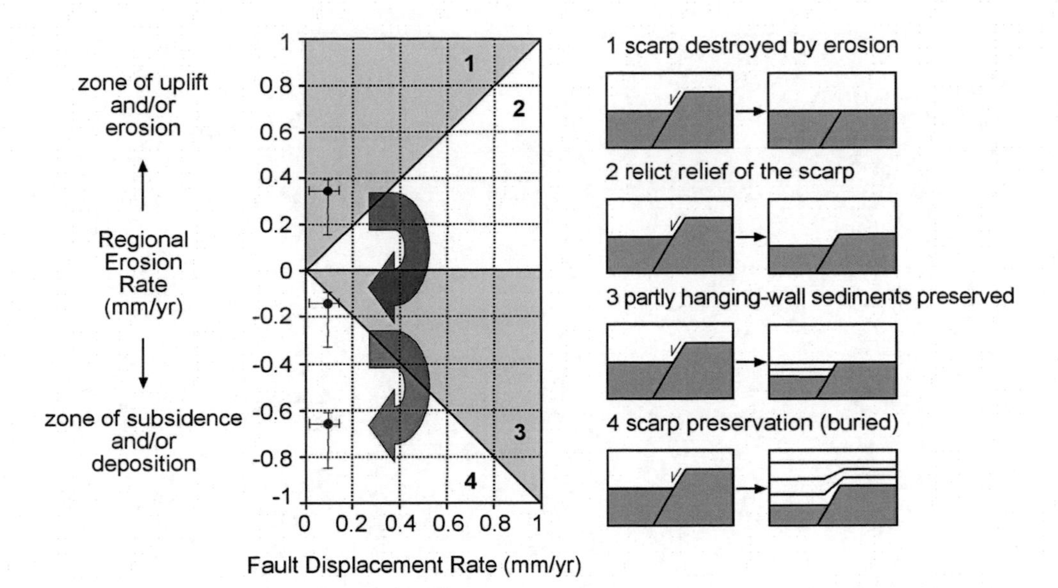

Figure 6.4.: Plot of fault displacement rate vs. erosion/deposition rate and related scarp evolution. A given constant displacement rate on normal faults leads, under different erosion (uplift) or deposition (subsidence) rates to the destruction of a fault scarp (sector 1) or to preservation of a topographic step due to a normal fault (sector 2). In sector 3 the fault scarp is buried by enhanced sedimentation. Blue and red arrows show effects if sedimentation rates are accelerated (sector 4). For further explanations see text. From Reicherter et al. (2011).

faults towards the center of the basin and within the lake.

Fault structures can not only be traced on land but also offshore within the lake, where steeply dipping normal faults constitute the margin of the lake as depicted on seismic and hydroacoustic profiles (Lindhorst et al., 2012a; Wagner et al., 2008). Offshore seismic data from Lindhorst et al. (2010, 2012b) show that the basement builds a deep central basin bordered by normal faults that are arranged in a rhombus shape. The steep eastern and western flanks distribute distinct morphological steps that give evidence of large N-S trending normal faults. Additionally, mass movement bodies found within the lake and also onshore (rockfalls, land-slides, sub-aquatic slides, homogenites, turbidites) are likely to have been seismically triggered (Lindhorst et al., 2012b).

Some of the observed fault scarps on land are exposed due to human activity through excavation of the fault breccia zone for construction material (fig. 6.5C) or by road and railway construction (fig. 6.5A, E).

In addition, the preservation of fault scarps is highly dependent on the exposition age; scarps in the heights of the Galicica Mountain range form only relicts (fig. 6.5B). Whereas scarps, which are located at lower heights along the Galicica mountain range (e.g. fig. 6.5 C) show nice striations and corrugations on the fault plane, as well as kinematic indicators like riedel shears (fig. 6.5E) or extensional comb fractures (tension gashes; fig. 6.5G; Angelier, 1994; Doblas, 1998). Generally, ophiolithic rocks like serpentinites have more, but smaller, fault scarps, with varying strike orientation. However, major N-S trending scarps mostly have a clear striation (fig. 6.5A, D). The slope of Galicica Mountain is characterised by stepped scarps (figs. 6.2A,E and 6.5F), forming a staircase-like landscape, which reaches down to the shore of Lake Ohrid and extends even further into the basin;

Figure 6.5.: Field photographs of geomorphological features. A: Striation (arrows) points to oblique slip on a NS trending normal fault with a dextral strike-slip component in serpentinitic rocks of the Mirdita Ophiolitic Zone located on the western shore south of Lini Peninsula (41°0′13.637"N, 20°38′1.097"E). B: Degraded relict fault scarp in the Triassic limestones (Korabi Unit) of the Galicica Mountain range (40°56′24.59"N, 20°48′53.478"E); dashed circle shows a person for scale. C: Excavated fault plane in Triassic limestones (L = lineation, arrow indicates normal movement) located at 40°58′8.292"N, 20°49′12.396"E. D: Steps on normal fault plane (arrow) in serpentinite, fault-bound shear lenses of Mirdita Ophiolite within the Korabi Unit; location: 41°3′55.508"N, 20°48′30.83"E. E: Polished normal fault with corrugations and rough parts in Triassic limestones of the Korabi Unit (arrow indicates movement of hanging wall, R = Riedel shear); location: 40°57′3.193"N, 20°48′1.274"E. F: View from Pestani (east coast) towards the N, note staircase-shape of the hill slope; this represents stepped tectonic scarps of normal faults. G: Scarp in Triassic limestones near Ljubanista(Korabi Unit) with horizontal extensional fractures, upper portion is degraded (above dashed red line), lower portion is fresh. Total scarp height is 2.5 m (40°55′7.331"N, 20°46′24.755"E). From Reicherter et al. (2011).

this has been described from high-resolution seismic data offshore (Lindhorst et al., 2012b; Wagner et al., 2008). Some fault scarps have been degraded by erosion in the upper portion of the free face, whereas the lower portion shows a uniform and planar free face testifying to a rapid exhumation of the fault plane (fig. 6.5G).

A T-LiDAR image of an excavated fault scarp at Galicica Mountains along the pass road (40°57′3.19" N, 20°48′1.27" E) shows asperities on the plane (grooves, lineations and corrugations as well as undulations, fig. 6.6). The white parts of the image are due to vegetation cover (fig. 6.6A). The N-S trending normal fault scarp, which is situated c. 980 m above sea level, has been scanned with 8-byte grey value distribution and 5 mm resolution (point by point). Based on the scan, a surface model was created exhibiting further details of the fault plane, including break-offs and riedel shears (fig. 6.6B). A stereographic projection (Schmidt, lower hemisphere) was created by measuring the orientation of the individual planes constituting the fault plane by laser scanning. Dips vary between 58° and 78° towards the west (fig. 6.6C). As a result, T-LiDAR scans help to obtain information on the fault plane (i.e. dip, dip angle and dip direction variation, undulation of the fault plane, asperities, lineation; for more information on LiDAR scans in the Ohrid Basin see Fuhrmann, 2009) and also help to reveal kinematic information when the indicators are well developed (e.g., riedel shears, steps, asymmetric steps, trails, fractures, see Doblas, 1998, for the variety of shear indicators on fault planes). Also LiDAR allows to obtain information from areas which are hard to access in the field, dangerous areas or quarries in operation.

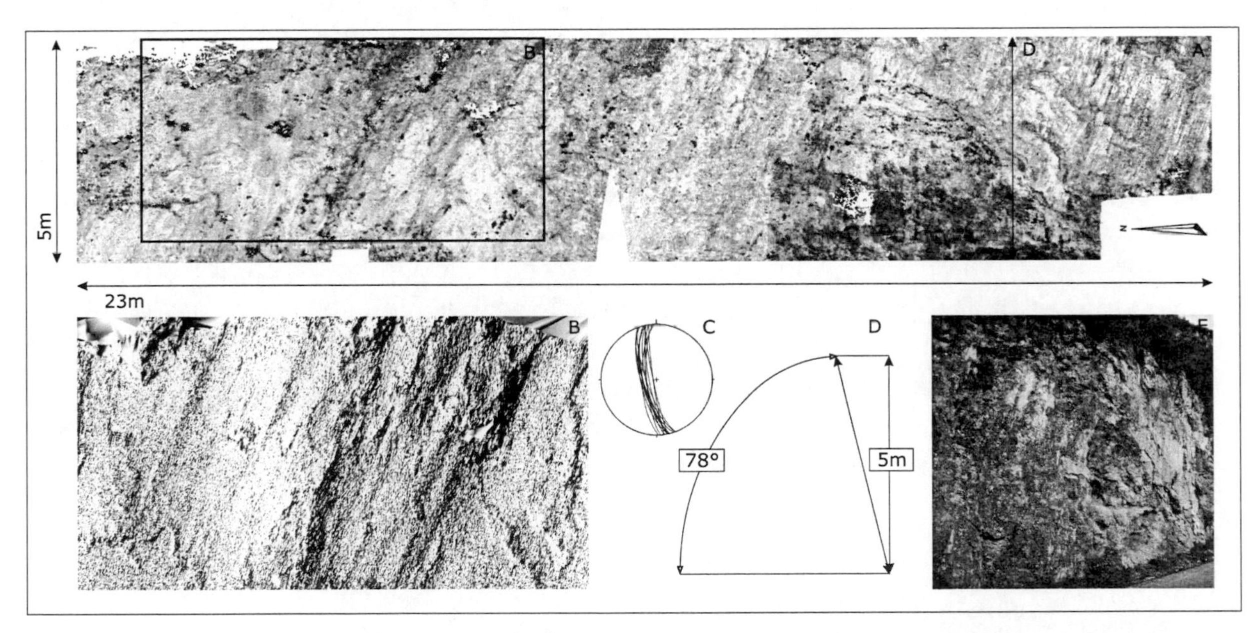

Figure 6.6.: T-LiDAR image of a fault scarp at Galicica Mountains along the pass road (40°57′3.19" N, 20°48′1.27" E; see also figure 6.5E) around 980 m above sea level. A: 23 m long N-S trending normal fault scarp has been imaged with 8-byte grey value distribution and 5 mm resolution (point by point), N is to the left. B: Surface model of a part of the scarp indicated in (A). C: Stereographic projection (Schmidt, lower hemisphere) of planes measured on the fault plane, dips between 58° and max. 78° are observed. D: Diagram showing the geometrical relationships between fault plane, height and dip angle. E: Photograph of the scanned section. From Reicherter et al. (2011).

To classify the seismic activity of an area and to gain a detailed picture of the landscape response to neotectonic movements the basin was investigated by detailed morphological studies (see chapter 2 for methods).

We measured profiles at over 30 sites (see fig. 6.1) consisting of different lithologies all over the basin to study the evolution of the basin's morphology and its response to tectonic forces. This was accompanied by various remote sensing techniques to provide the link between landscape evolution and tectonic activity in the region.

6.1. Morphological markers

6.1.1. Triangular facets

Along both the eastern and the western coastlines abundant triangular facets with hanging valleys or wine-glass shaped valleys, which dissect the faceted spur ridges, are found (fig. 6.7). According to the hypothesis of Benedetti et al. (2002) and Papanikolaou et al. (2005) which follow the ideas of Mercier et al. (1976), Stewart and Hancock (1990) or Vittori et al. (1991), the fault scarps began to develop after the Late Glacial Maximum in the wider Mediterranean region. The observed hanging valleys, which are related to active faults, are therefore, important landforms along the N-S running borders of Lake Ohrid. Triangular facets are the result of normal faulting along a mountain ridge. The base level fall and incision of streams divides the hillslope into several drainage basins resulting in the triangular shape of the mountainfront. Typical for active triangular facets systems is a straight mountain piedmont junction and narrow valleys. With the next faulting event a new free face is created and the valleys are interrupted and thus become hanging valleys. If streams and valleys are hanging, sometimes waterfalls develop (also called wine-glass shaped valleys); however, most of the rivers at Lake Ohrid are dry and yield only small quantities of seasonal water. In times of quiescence the mountain fronts are building up sinuosity with the retreat of the range bounding fault and the valleys become wider. The initial streams develop and become small elongated watersheds. The expression of triangular facets is also highly connected to the erosional resistance of rocks and the climatic conditions as shown by Bull (2007). Schematically, the generation of triangular facets and associated wine-glass shaped/hanging valleys is depicted in figure 6.7.

In the Lake Ohrid area triangular facets can be found at two locations (see fig. 6.8). One is in the southwest above Hudenisht (fig. 6.8A), the second is in the north close to the villages of Trebenista and Gorenci (fig. 6.8C).

Along the western shore of Lake Ohrid, repeated triangular facets (figs. 6.2D and 6.7A) originate in stepped normal faults. The heights of the facets range between 162 m and 210 m and have a length at the mountainfront piedmont junction between 570 m and 750 m. The river courses are partly deflected testifying to parallel normal fault activity with a certain oblique component, e.g. near Hudenisht (figs. 6.8A and 6.2E) where four normal faults form a landscape with a stepped escarpment. The mountainfront sinuosity indices calculated for the Hudenisht area provide evidence for the relativeley young age of the triangular facets; the material along the south western coast is relatively easier to erode. For the southern part of the Hudenisht area a mountainfront sinuosity index of 1.091 was calculated and a dissection stage between 2 and 3 was determined (after Bull, 2007). For the northern part of the Hudenisht area, a mountainfront sinuosity index of 1.075 and a dissection stage of 5 was calculated (after Bull, 2007). Furthermore, in front of the mountain range in the southwestern part of Lake Ohrid, inactive alluvial fans form a linear coastline instead of a lobe shaped fan complex (fig. 6.2E). The shape of the alluvial fans with the linear cut edge could be interpreted as drowned fans by Holocene lake level rise. However, high resolution hydroacoustic surveys demonstrated shoreline parallel active faults offshore (Lindhorst et al., 2010) which is more likely the reason for the cut off fans.

Triangular facets in the north of Lake Ohrid are more heavily eroded and exhibit large alluvial fans in between the faceted spurs. The heights of the facets here range between 162 m and 250 m and have a length at the mountainfront piedmont junction between 750 m and 1200 m. A mountainfront sinuosity index of 1.13 was calculated here, although the area seems to be a lot less active than the Hudenisht area and a dissection stage between 3 and 4 was determined (after Bull, 2007). Values of the mountainfront sinuosity index between 1.0 and 1.5 reflect a highly active mountainfront after Bull (2007).

The range bounding faults at both locations trend NW-SE and NE-SW but follow in general the N-S trend of the youngest deformation phase described in chapter 4. They are therefore activated or reactived by the E-W trending extensional regime dominating the basin since Holocene.

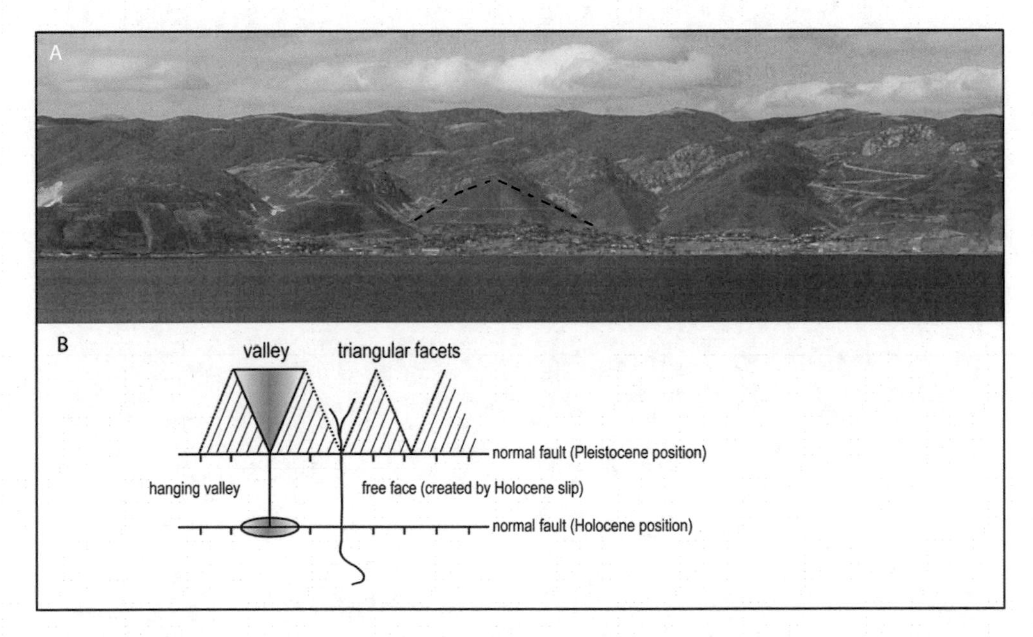

Figure 6.7.: A: Triangular facets along the west coast of Lake Ohrid near Hudenisht village (Albania). In front of the range the alluvial fans are fault-cut and form a linear coastline. B: Schematic evolution of triangular facets and wine-glass shaped or hanging valleys regarding the changes from Pleistocene to Holocene climatic conditions. From Reicherter et al. (2011).

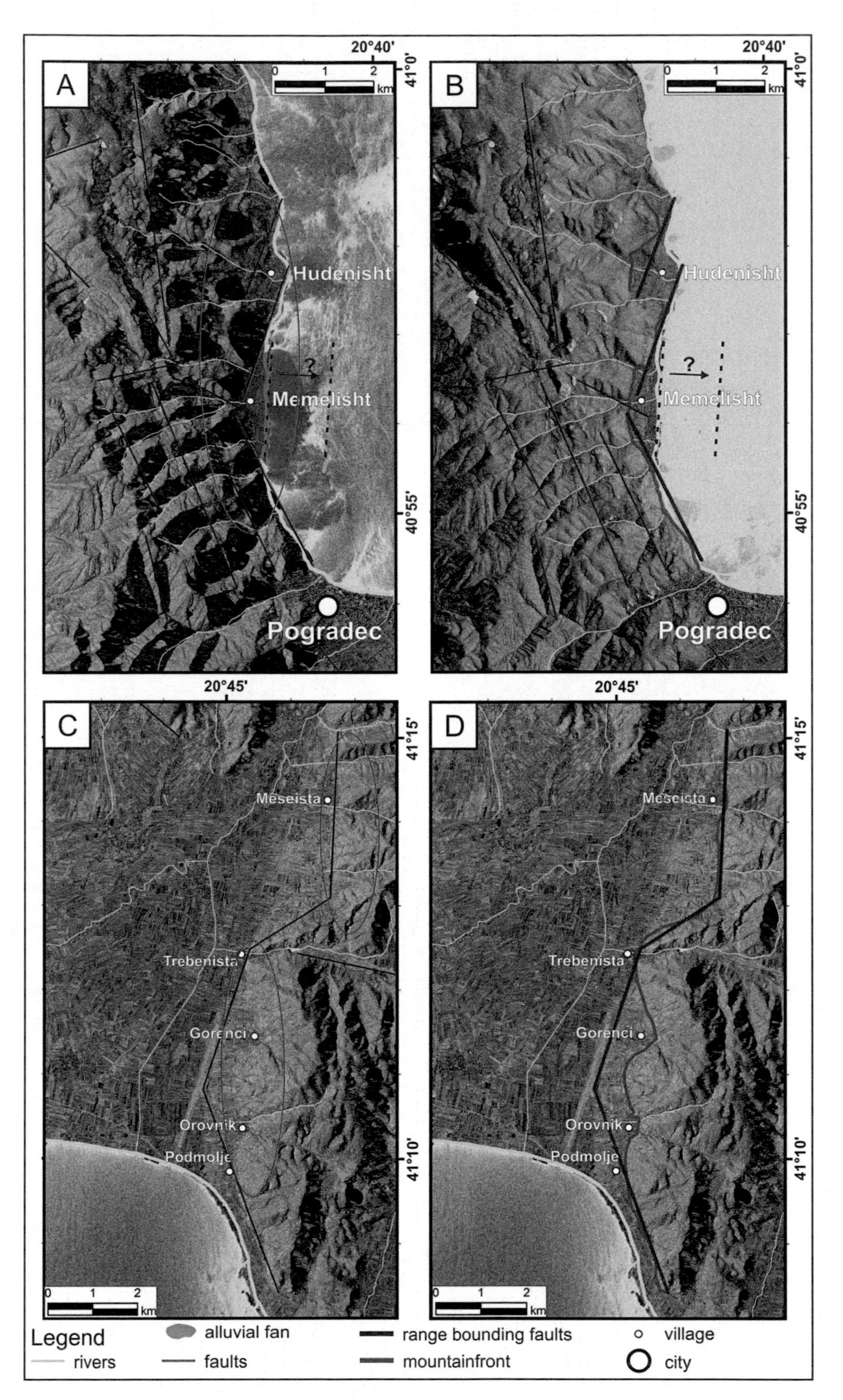

Figure 6.8.: Mountainfront Sinuosity Index for two sites at Lake Ohrid, both exposing triangular facets in various states of degradation. A: Map view showing main faults in the Hudenisht area. Yellow ellipse shows location of triangular facets and associated alluvial fans. B: Mountainfront and range bounding faults in the Hudenisht area. C: Map view showing main faults in the Meseista and Gorenci area. Yellow ellipse shows location of triangular facets and associated alluvial fans. D: Mountainfront and range bounding faults in the Meseista and Gorenci area.

6.1.2. Wind gap

A major wind gap forms the pass between Lake Ohrid and Lake Prespa, which separates the Galicica Mountains from Mali I Thate Mountains (fig. 6.2C) and may be a very old landscape feature. E-W directed faults are rare in the observed region (e.g., figs. 3.10 and 6.8). Lindhorst et al. (2012b) also did not detect any E-W trending faults in the seismic sections and bathymetric data. The wind gap between Lake Ohrid and Lake Prespa does not have a clear counterpart at the western lakeshore. The prolongation is possibly the E-W trending valley close to the Albanian-Macedonian border at the Lini Peninsula with a related polje in the valley of Perrenjas (fig. 6.1). Another E-W trending fault is mapped by Aliaj et al. (1995) in the Hudenisht area. Along the wind gap the blocks of Galicica and Mali I Thate Mountains are rotated so that the strike of both varies by about 20°. The western slope of the Galicica range is bound by repeated fault scarps accompanied by impressive hanging valleys (fig. 6.9).

We interpret the pass as a remnant valley, linking Lake Prespa with Lake Ohrid. It is most propably the result of the Eocene to Oligocene stress field (see chapter 4) that developed also E-W trending transtensional basins in eastern Macedonia. Below the wind gap a bowl shaped polje structure preserves large amounts of carbonate-cemented colluvial sediments as mushroom shaped structures. The river mouth was once located south of the village of Trpejca, where today a little creek enters the lake (see figs. 3.14B, and 6.9E, F). The angular shaped grains point to a very short transport distance and most likely originate from the slopes of the Galicica. Today water transport into Lake Ohrid is mainly via karstic caves and springs which account for most of the influx from Lake Prespa (Albrecht and Wilke, 2009; Matzinger et al., 2006a, see chapter 3.3.4), surface run-off only contributes to water inflow to a small extent. The creek at Trpejca has also lost its transport capacity compared to the amount of transported gravel visible in the described outcrop above.

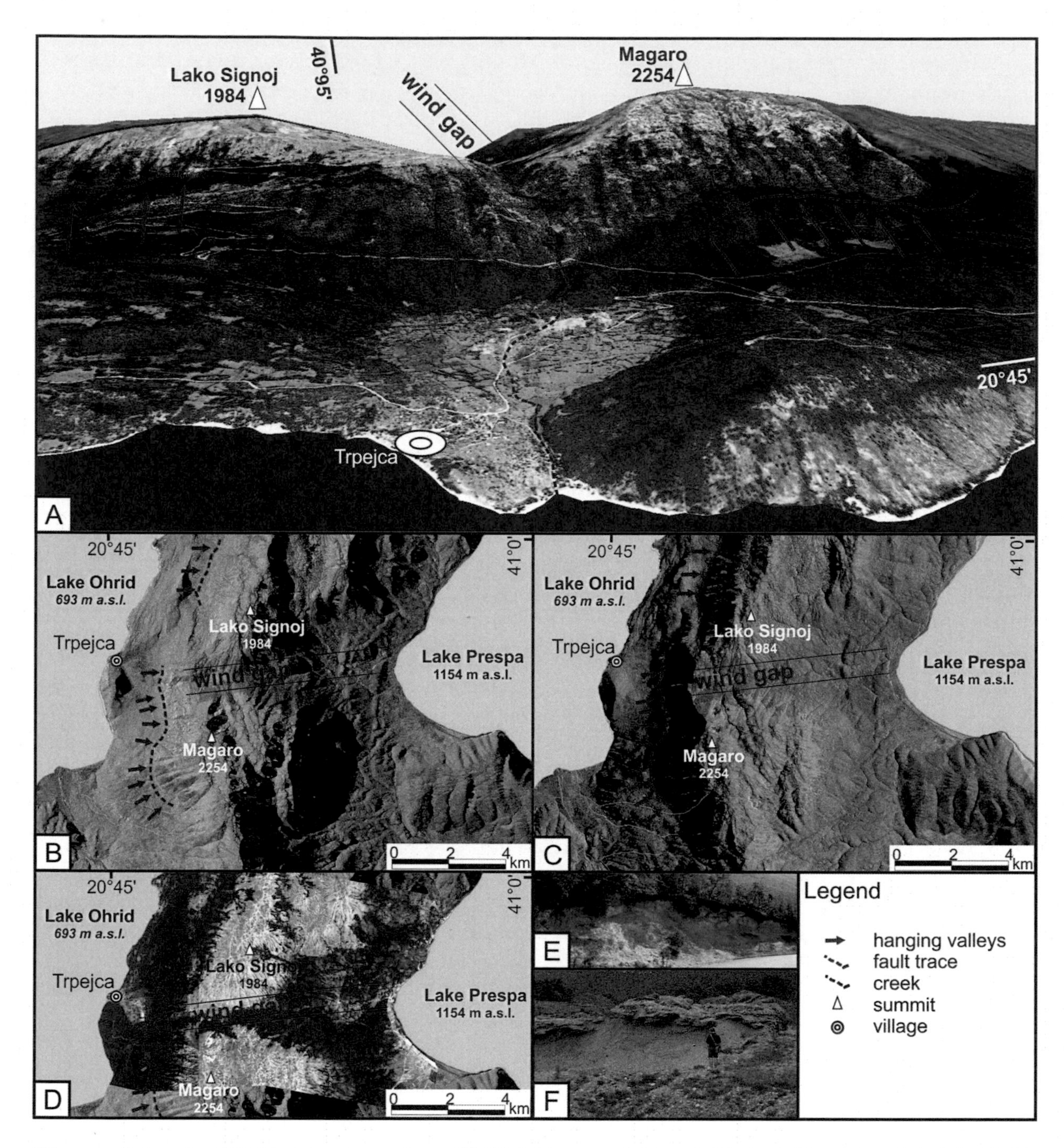

Figure 6.9.: A: SPOT data draped on SRTM (1 arcsec) based Digital Elevation Model. B-C: Map view of the wind gap and hanging valleys (marked by red arrows) at the southeastern tip of Lake Ohrid. Red dashed lines indicate fault trace. B: Scan with a 73° angle flying south. C: Scan with a 82° scanning angle flying south. D: SPOT data draped on TerraSAR-X mosaic. E: Carbonate-cemented colluvial sediments preserved in an outcrop along the Galicica road.
F: Carbonate-cemented colluvial sediments preserved in a mushroom like shape.

6.2. Scarp profiles

Bedrock fault scarps are the most visible land forming feature, that are observed in the Ohrid region and they account for a high percentage for the morphological expressions within the basin. As previously mentioned, they are considered to be postglacial, as the slip-rate along the fault planes needs to be higher than the erosion-rate, to preserve the step-like morphology. Bedrock fault scarps are found in all lithologies around Lake Ohrid and are well preserved in limestones and ophiolites. In weaker rocks, which are easily erodible, such as serpentinite or unconsolidated sandstones, fault scarps are generally strongly degraded. An example is shown in figure 6.10 with a serpentinite fault scarp located directly below the Boces fault scarp (fig. 6.20). Here, the principal shape of a fault scarp is given but it is significantly smoothed by erosion. These scarps were not considered for profile quantifications, as they are scarce and not comparable to the more common resistant rocks. They are, however, interesting when it comes to regional classification of activity. 31 fault scarps were measured throughout the basin; 14 along the western graben shoulder, 14 along the eastern graben shoulder and 3 in the north of the basin. Here, a few examples are presented in detail. The denomination of the scarps sections is explained in chapter 2.5.2.

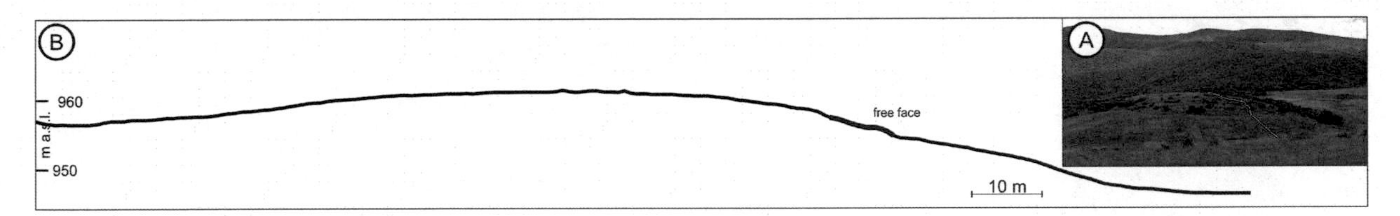

Figure 6.10.: A: Photograph of the Boces serpentinite scarp. Yellow line gives the approximate position of the profile. B: Complete profile of the Boces serpentinite fault ($41°02'51''N, 20°37'31''E$; 967 m a.s.l. for location see figure 6.1). No vertical exaggeration.

6.2.1. East coast

Koritsi Rid

The location Koritsi Rid is used as a touristic viewpoint (for location see figure 6.1). Therefore, the main, and actually better preserved, scarp could not be measured due to anthropogenic modifications (e.g. a parking space above the scarp, anchorage for a fence). Another limestone scarp located a few metres north of the main scarp has been profiled and exhibits a free face with a offset of about 23 m (fig. 6.11). The fault plane dips to the west with an overall angle of 62°, the lower slope has a steep gradient of 38°. Very little colluvium has built up. What little there is is comprised of loose limestone blocks (max. 30 cm in diameter) found spread on top of the bedrock with soil located in between the blocks. In general the vegetation cover is not intense because of grazing animals (mainly sheep), which are kept in the area. Above the free face the upper slope is degraded, whereas the top of the free face is well preserved. A low angle bench halfway down the free face is covered with meadow and dips southwest. Below the step, the free face is strongly degraded and not as clearly defined as the upper part. The upper slope

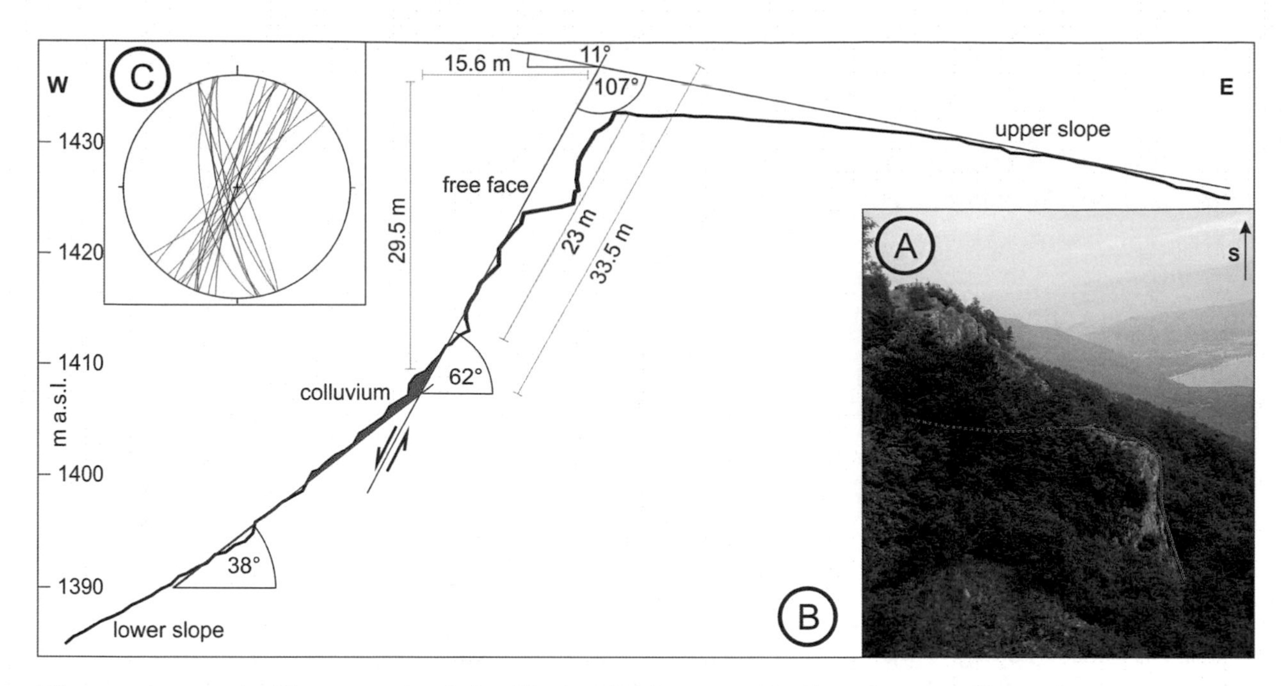

Figure 6.11.: A: Photograph of the Koritsi Rid scarp. B: Complete profile and geometric reconstruction of the Koritsi Rid fault ($40°57'55''N, 20°48'40''E$; 1425 m a.s.l., for location see figure 6.1). No vertical exaggeration. Green lines show constructed angles of the upper and lower slope and the fault/joint plane. Red lines indicate the free face measured in the field. The colluvium (marked in yellow) is constructed from field data. C: Stereonet plot of measurements from the fault/joint plane.

is rotated 11° clockwise from horizontal, which means that if the block is rotated back to horizontal, the original dip of the fault plane would at least have been 73°; this is still low compared to other scarps in the same lithology. Considering only the upper part of the free face the measured angle is at least 82°. Adding the rotation of 11°, a free face angle of 93° would result which is unlikely. The bench is interpreted either as a joint taking up stress, or as the bottom joint of a detached block that is now located somewhere below the scarp. This is supported by the well preserved top of the free face compared to the rest of the scarp, which is strongly degraded. This leads to the assumption that the top part is younger (or exposed to erosion later) than the bottom part. Although being younger and better preserved, the plane does not show any striation that would qualify it as a fault plane. So that in this case the plane does not represent the true scarps free face, but a joint plane being parallel or subparallel to the fault plane. Together with the rotation problem this underlines the idea of a detached block.

Lako Signoj

The Lako Signoj limestone scarp is composed of two single steps (for location see figure 6.1). The upper scarp shows a offset of about 11 m, the lower one about 14 m (fig. 6.12). The most visible difference between the two is that the horizontal displacement significantly varies (2.3 m upper scarp and 9.5 m lower scarp) while the vertical displacement is relatively uniform at around 11 m. The dip angle of the upper fault is 85.5°; it is quite normal for these limestone faults to be very steep and show angles close to vertical. In contrast, the dip angle of the lower fault is only 54°. The upper slope is heavily degraded close to the edge of the fault. Therefore, the upper edge of the fault can not clearly be defined; there might have been a detached block at the top giving it its distinct shape. The volume of colluvium between both scarps is very small compared to the volume of the upper slope's missing material. The 25.5° slope gradient is measured for all three slopes. The lower scarp seems to be less eroded and exhibits even smaller volumes of colluvium. The ground cover directly below the lower scarp is composed of some tens of cm of soil. Along the relatively steep slopes the eroded material has been quickly transported further downslope; this will especially happen at certain times of the year, for example, with the beginning of snowmelt in spring. The reconstructed slip of both scarps is about 15 m each.

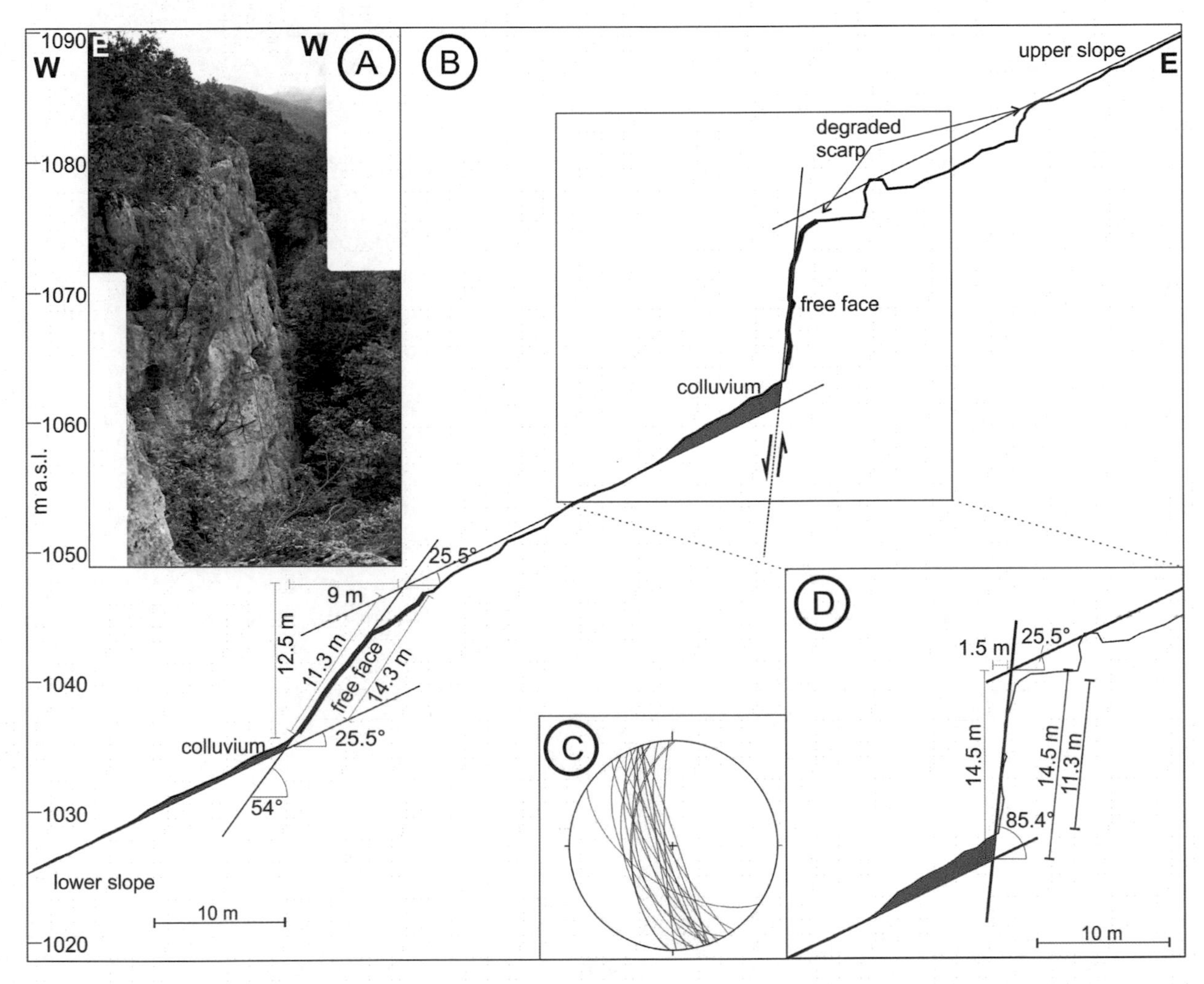

Figure 6.12.: A: Photograph of the Lako Signoj scarp. Yellow line gives the approximate position of the profile. B: Complete profile and geometric reconstruction of the Lako Signoj fault ($40°58'11''N, 20°48'07''E$; 1076 m a.s.l. for location see figure 6.1). Only the upper scarp was analysed in detail. No vertical exaggeration. Green lines show constructed angles of upper and lower slope and the fault plane. Red lines indicate the free face measured in the field. The colluvium (marked in yellow) is constructed from field data. C: Stereonet plot of measurements from the fault plane. D: Dip angles and offset derived from geometric construction.

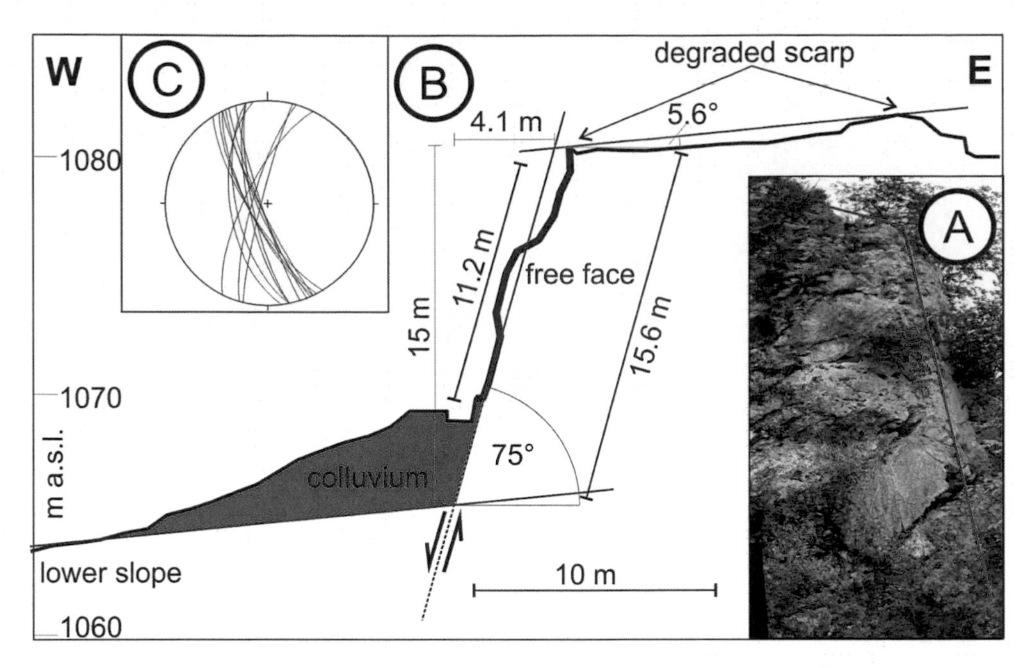

Figure 6.13.: A: Photograph of the Elsani scarp. Yellow line gives the approximate position of the profile. B: Complete profile and geometric reconstruction of the Elsani fault ($41°01'21''N, 20°49'19''E$; 1076 m a.s.l. for location see figure 6.1). No vertical exaggeration. Green lines show constructed angles of upper and lower slope and the fault plane. Red lines indicate the free face measured in the field. The colluvium (marked in yellow) is constructed from field data. C: Stereonet plot of measurements from the fault plane.

Elsani 1

Elsani 1 is one of the smaller scarps with only 11 m of measurable offset (for location see figure 6.1). The upper and lower slopes are very flat at 5.6°, which is probably due to the proximity to the shoreline (fig. 6.13). The fault plane has a dip angle of 75°. The colluvium is well formed but modified by a military trench directly below the free face. The reconstructed offset is 15.6 m with an extensional component of 4.1 m.

Dolno Konjsko

Above the Metropol Hotel at the village of Dolno Konjsko a (fig. 3.11) a palaeosol is preserved in the hanging wall of an active normal fault (figs. 3.14C and 6.14), where Quaternary sediments are displaced against Jurassic limestones. The palaeosol was dragged into the fault during the fault movement and is, therefore, a marker for activity in the basin. OSL dating (see chapter 2.4.2 for details) was performed at this location to get a minimum age of the last movement along this fault. The probing locations are marked in figure 6.14. Only the sample of the stratigraphically youngest sediment delivered reproducible results, as the OSL signal was already in saturation for samples OH2, OH3 and OH4 and therefore did not give an age determination.

The OH1 sample delivered a Pleistocene age dated to 11 ± 1.7 ka. This period can be assigned to the Younger Dryas stadial, which marks the transition between Pleistocene

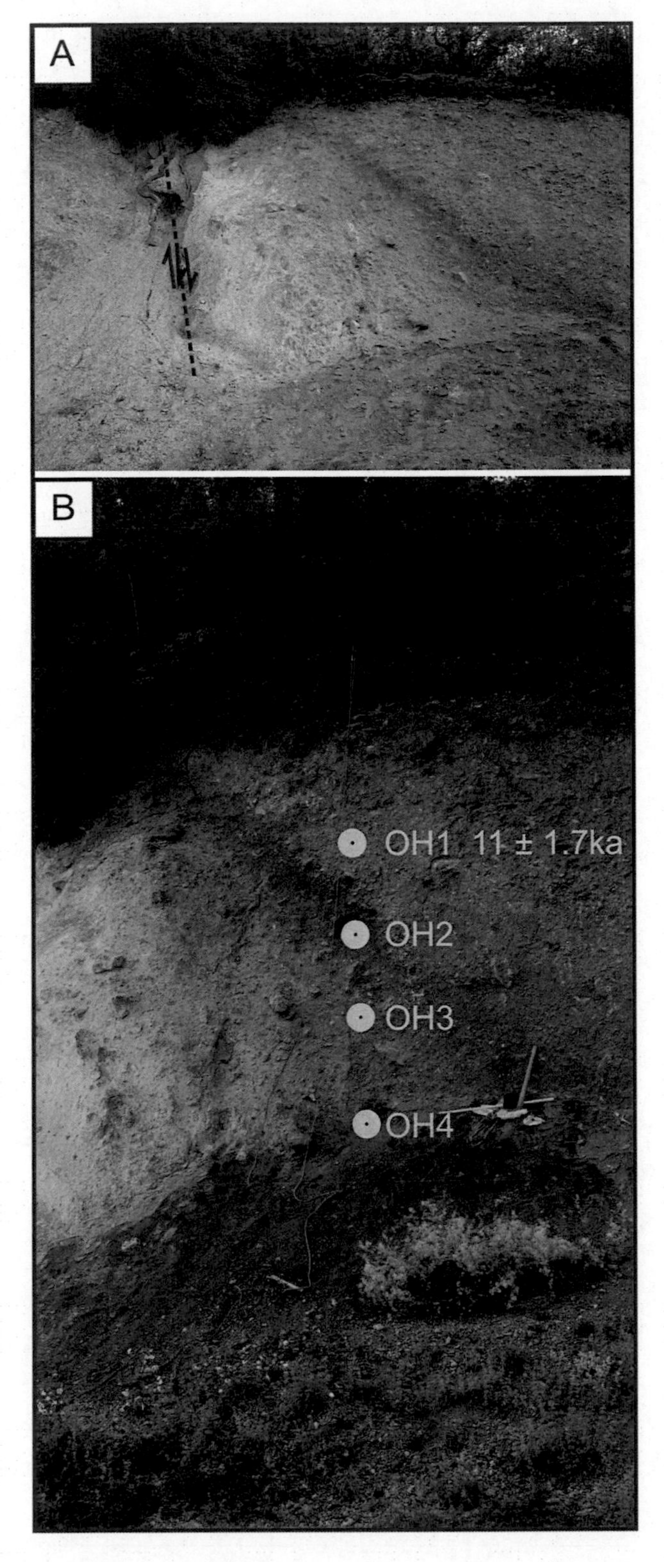

Figure 6.14.: A: Fault at Dolno Konsjko where Quaternary sediments are displaced against limestones. B: OSL sampling locations and associated age. For location see fig. 6.1.

and Holocene. The material deposited here is composed of unrounded large limestone clasts originating from the Triassic limestones of the Galicica mountains. Erosion and transportation mechanisms are favoured by the scarce vegetation and dry soils (Vogel et al., 2010a). Below, the palaeosol is characterised by red coloured soil with root remnants and much smaller grain sizes than the clasts above and below. This portion is ascribed to the Alleröd, which represents the last interstadial with warmer climate conditions allowing soil formation. Vogel et al. (2010a) found evidence for cold winters and higher spring-summer temperatures in offshore sediment cores. Below again a section containing large limestone clasts occurs pointing to colder conditions and high erosion rates as present in the Older Dryas.

The outcrop can be therefore divided into 4 sections:

- recent soil in red colour (Holocene)
- upper part with large clasts; light brown colour (sample OH1; Younger Dryas)
- red coloured, dragged palaeosol and plant remnants, roots (sample OH2; Alleröd interstadial)
- lower part with large clasts; light brown color (sample OH3 and OH4; Older Dryas)

This outcrop shows that normal faulting processes took place already during Pleistocene times. But erosion erased any geomorphological evidence of faulting (e.g. fault scarps; fig 6.3A) so that the sense of slip and the orientation of the fault can only be estimated by the dragged palaeosol. This leads to the assumption that the fault scarps exposed around Lake Ohrid must at least be younger than the Alleröd interstadial or 11 ka respectively.

6.2.2. West coast

Sveti Arhangel 1

Two profiles have been constructed for the Sveti Arhangel fault, both showing almost the same expression; therefore, only one is presented here (for location see figure 6.1). Sveti Arhangel 1 is one of the highest faults in our studied set with almost 60 m of offset (fig. 6.15). It is not clear whether this fault has been exposed by anthropogenic factors such as exhumation/excavation for road and housing construction.

The upper part of the scarp does not look very degraded; however, the cover is mainly made up of grass and a few small trees, which can not hold back a significant amount of material that comes from above. The cover, therefore, causes the colluvium to look a lot larger compared to the volume of material which has fallen from the upper degraded part of the scarp. The free face angles have been interpolated between measurements at the top and bottom of the free face as the whole free face was not accessible and is relatively steep at 81°. The accumulated colluvium has a gradient of 27.2°. Close to the shore the slope angle flattens almost to 0° and bedrock crops out. There are two possible explanations for the formation of this flat plain. On the one hand it may have been caused by human modifications; at some hundreds of metres away a deserted hotel is located, that might have had its beach area there. On the other hand, it may have been caused by the influence of the lake by erosion and lake high stands.

The Sveti Arhangel fault distributes a large riedel shear with a width of some tens of

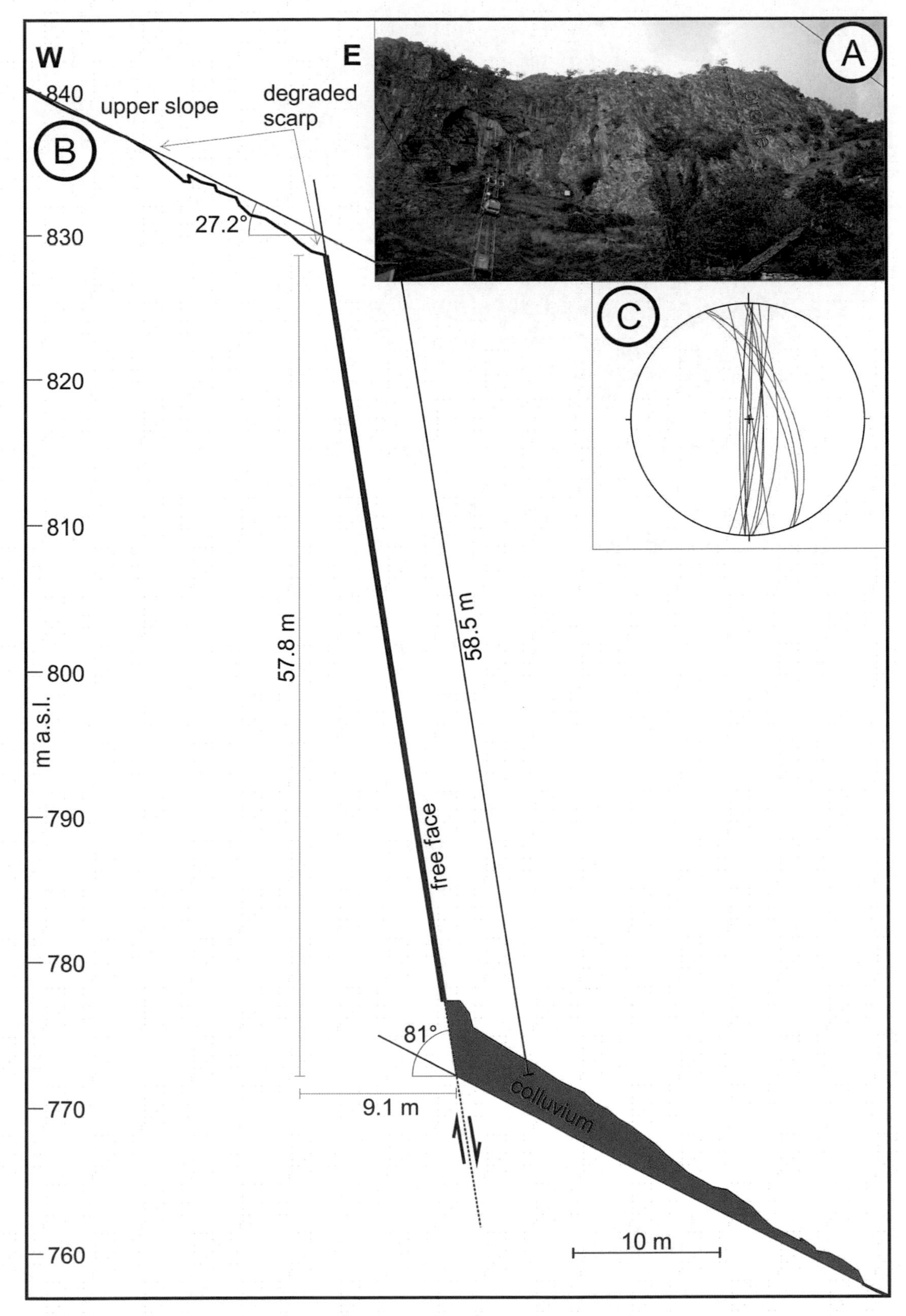

Figure 6.15.: A: Photograph of the Sveti Arhangel scarp. Yellow line gives the approximate position of the profile. B: Complete profile and geometric reconstruction of the Elsani fault ($41°06'27''N, 20°37'56''E$; 840 m a.s.l., for location see figure 6.1). An average dip was calculated from measurements at top and bottom of the scarp as the free face was not completely accessible. No vertical exaggeration. Green lines show constructed angles of the upper and lower slope and the fault plane. Red lines indicate the free face measured in the field. The colluvium (marked in yellow) is constructed from field data. C: Stereonet plot of measurements from the fault plane.

Figure 6.16.: A: Photograph of the Sveti Arhangel scarp looking north. Dashed red lines show almost vertical joints. B: Scarp with the Sveti Arhangel Mihail church located below a large riedel. Red arrows point to riedel location. C: Vertical joints/fault planes at Sveti Arhangel.

metres along the NNW-SSE trending fault (fig. 6.16B). This riedel was created by strong compressional forces in NE-SW direction as described in chapter 4 (Sv. Arhangel II; compressional stress state). During the compressional phase joints and faults formed parallel to the observed riedel (fig. 6.16A, C), which were reactivated by the following E-W extensional phase, which most likely led to the formation of the impressive fault scarp at Sveti Arhangel. Along this steeply dipping joints and secondary faults (no lineations could be measured as the limestone is coated with calcite), the fault planes/bedding planes are leaning against each other like dominoes whilst the exposed face is undergoing gravitational failure. This fault has, therefore, been reactivated along weakness zones. The actual bedding dips towards west (no data available).

The church Sveti Arhangel Mihail, which is built into the limestone wall (fig. 6.16B), dates back to the 13th century. As the church has been located on the fault plane for about 800 years and remains undamaged, the fault has not been active at least since then.

Hudenisht

The fault scarp above Hudenisht (for location see figure 6.1) is a limestone fault scarp (fig. 6.17) with a fee face dipping at 81°. The upper slope gradient has an angle of 5° westwards, whereas the lower slope gradient is 25° to the east. The fault scarp itself is strongly eroded on its top edge. The recent free face has 16.5 m of offset while the reconstructed fault scarp has a offset of 29.6 m. The colluvium below is relatively large and well defined. The gradients of upper and lower slopes are in this case not identical

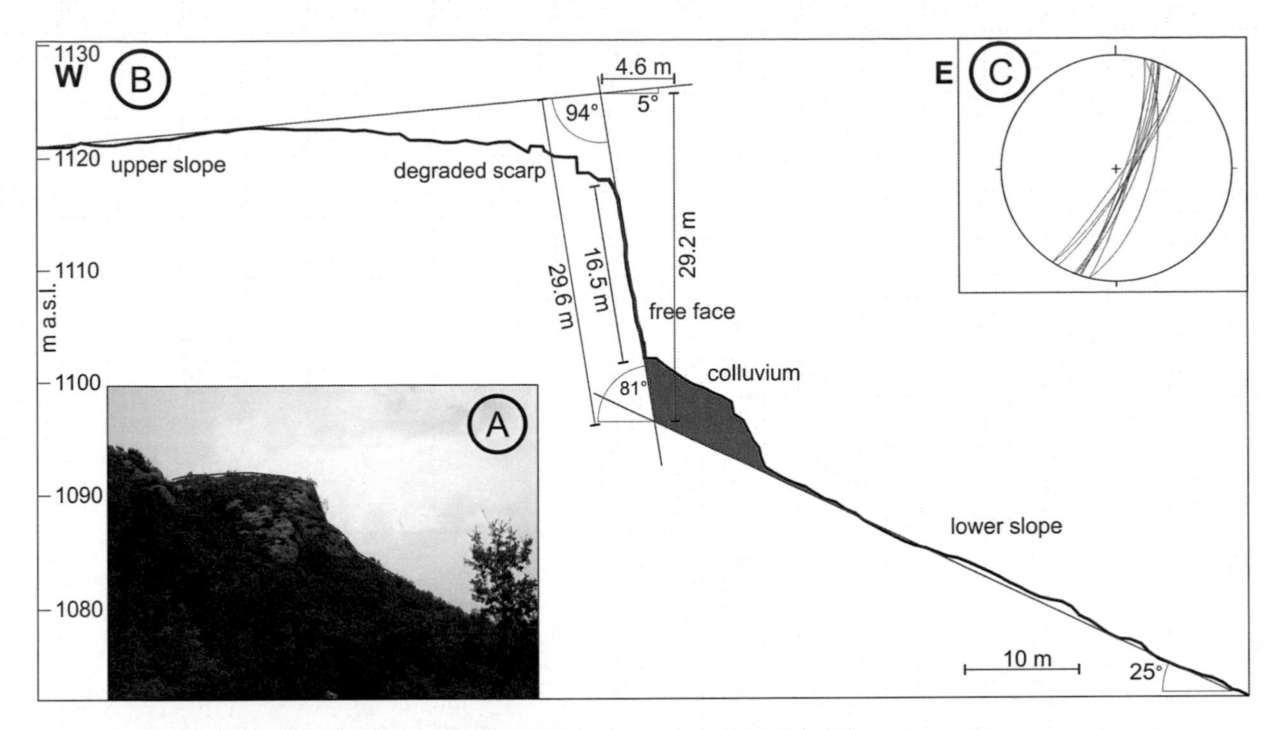

Figure 6.17.: A: Photograph of the Hudenisht scarp. Yellow line gives the approximate position of the profile. B: Complete profile and geometric reconstruction of the Hudenisht fault ($40°57'08''N, 20°37'43''E$; 905 m a.s.l., for location see figure 6.1). No vertical exaggeration. Green lines show constructed angles of upper and lower slope and the fault plane. Red lines indicate the free face measured in the field. The colluvium (marked in yellow) is constructed from field data. C: Stereonet plot of measurements from the fault plane.

but are even in opposite directions. If the block is rotated 5° anticlockwise, so that the upper slope is horizontal, the dip of the fault plane changes to 86°.

The Hudenisht fault scarp is a relatively simple example of a rotated block. The colluvium is made up of large blocks so that is was hard to distinguish it from the in-situ basement rock. However, during profile construction, the contact is clearly visible so that a potential mismatch is disregarded.

Piskupat

The Piskupat fault scarp (fig. 6.18) is somewhat more unique than the other faults found in the Lake Ohrid Basin (for location see figure 6.1); there are quite a few similar to this one on the west coast, but they were not accessible. The fault plane angle is 39° with well distributed colluvium below. The lower slope has a gradient 11° to the east. The upper slope, which is made up ophiolites, has a gradient of 14° towards east. This is slightly more than the lower slope, which is due to a different formation process (faulting versus erosion). The key feature is the change of lithology which causes the specific morphological shape. The upper slope is made up of ophiolites while the scarp itself and the lower slope is made up of limestones. The ophiolites have been eroded more quickly due to the different resistance to weathering. As a consequence the limestones are now exposed and preserved as a fault scarp (see fig. 6.19). The eroded material, that should

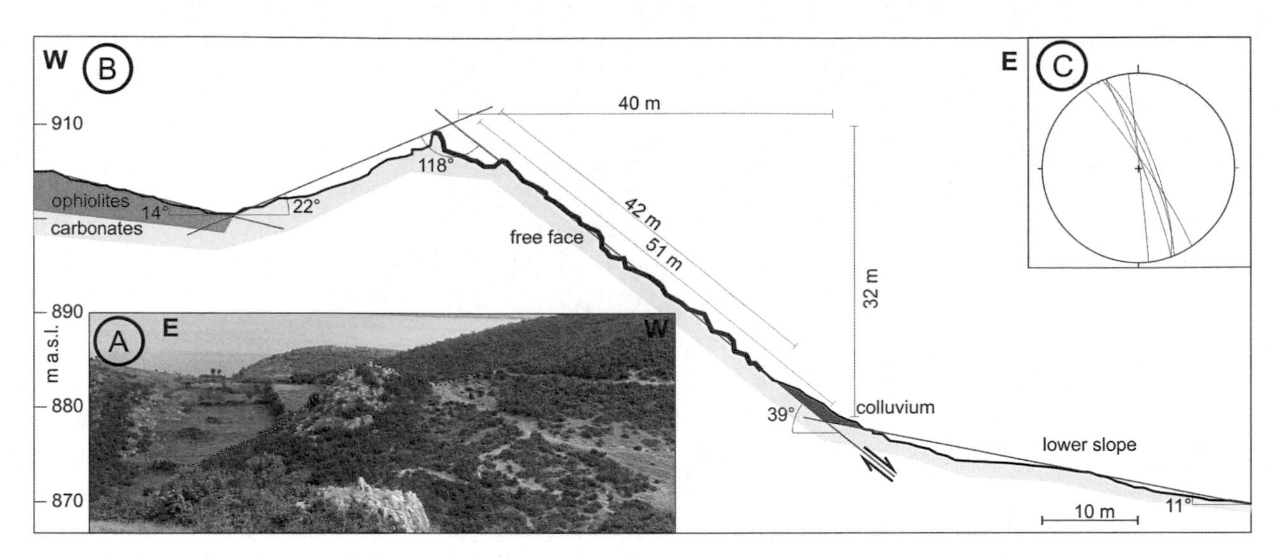

Figure 6.18.: A: Photograph of the Piskupat scarp. Yellow line gives the approximate position of the profile. B: Complete profile and geometric reconstruction of the Piskupat fault ($41°02'15''N, 20°37'32''E$; 905 m a.s.l., for location see figure 6.1). No vertical exaggeration. Green lines show constructed angles of upper and lower slope and the fault plane. Red lines indicate the free face measured in the field. The colluvium (marked in yellow) is constructed from field data. C: Stereonet plot of measurements from the fault plane.

be found in the halfgraben, can be transported to the north and into the lake via a small river valley.

The contact between ophiolites and limestones mimics the presence of a fault plane, if looking at the morphology of the scarp, only. But the actual fault plane is not located in the small halfgraben where the change of lithology was observed; it must be located further up to the west. This assumption considers erosional effects that cause the contact point of lithologies to migrate upslope (see fig. 6.19).

Boces

This fault profile is located at the village of Boces on the west coast (for location see figure 6.1). It shows a set of two fault scarps (fig. 6.20), that developed in the Upper Cretaceous limestone caps. Both scarps exhibit an angle of 123° between upper slope and associated fault plane. The upper scarp is rotated 15° anticlockwise compared to the lower scarp. From both scarp's unrotated positions, the clockwise rotation to horizontal is 2.5°. If all rotation were undone, the upper fault plane would have about the same dip (57°) as the lower fault plane (57.5°). The lower slope is relatively steep with an angle of 30.8°. The volume of colluvium here is smaller than the upper colluvium (between both scarps).

From the field pictures and from SRTM data, it can be depicted that the transition between slope and the coastal plain is abrupt and very clearly defined. The free face of the lower scarp shows 30.5 m of offset (see table 6.1) while the upper scarp has a offset of 27.6 m. The reconstruction of the original scarp's shape shows that the upper scarp is

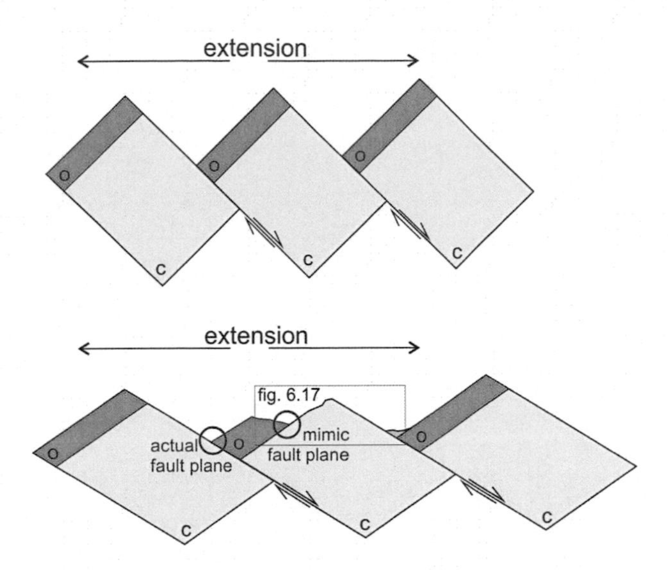

Figure 6.19.: Concept of fault scarp evolution at the Piskupat fault. **O**=ophiolite, **C**=carbonate. Planar rotational faults evolve under an extensional stress regime. Later these are subject to erosion. As the erosion strips off the ophiolites from the underlying carbonates the shape of the fault can evolve. The halfgraben below the fault is filled with debris from the surrounding area.

more eroded on its top edge while the lower scarp shows only little degradation at the top. Stronger degradation and stronger rotation suggests that the upper scarp is older and has therefore undergone more deformational phases than the lower one. This might be due to ongoing faulting and consequently further rotation along the fault plane. This is also evidenced by the stronger degradation at the top edge of the upper scarp compared to the lower one. After profile reconstruction, the upper scarp shows an offset that is higher (36.54 m) than the offset at the lower scarp (32.66 m). The colluvium located in between both scarps is thicker because the material coming from above is deposited in a heavily eroded halfgraben, which better preserves the debris compared to the steep slope below the lower scarp. Here, the material is quickly and easily transported downhill and directed into a small neighbouring river. This process allows the colluvium to spread out more widely and not accumulate into a typical wedge shape.

6.2.3. North

Delogozdi 2

The Delogozdi 2 scarp is one of three profiles from the north of the basin. The reason why very few fault scarps are preserved north of the lake is that the lithology is composed of mainly phyllites, which are not very resistant to erosion. Therefore, only faults located in limestones are present in the landscape (for location see figure 6.1). This very long profile shows a fault scarp with a free face height of 6 m and a offset of almost 7 m (fig. 6.21). The fault plane is not very steep at only 61.5°. It mainly dips west with variations of c. 20°. The upper and lower slopes have an angle of 21°. The reconstruction of the profile delivers a different picture. In the reconstruction the height of the scarp is 18 m while the offset is almost 21 m. Most of this is filled up with colluvium today. Degradation of

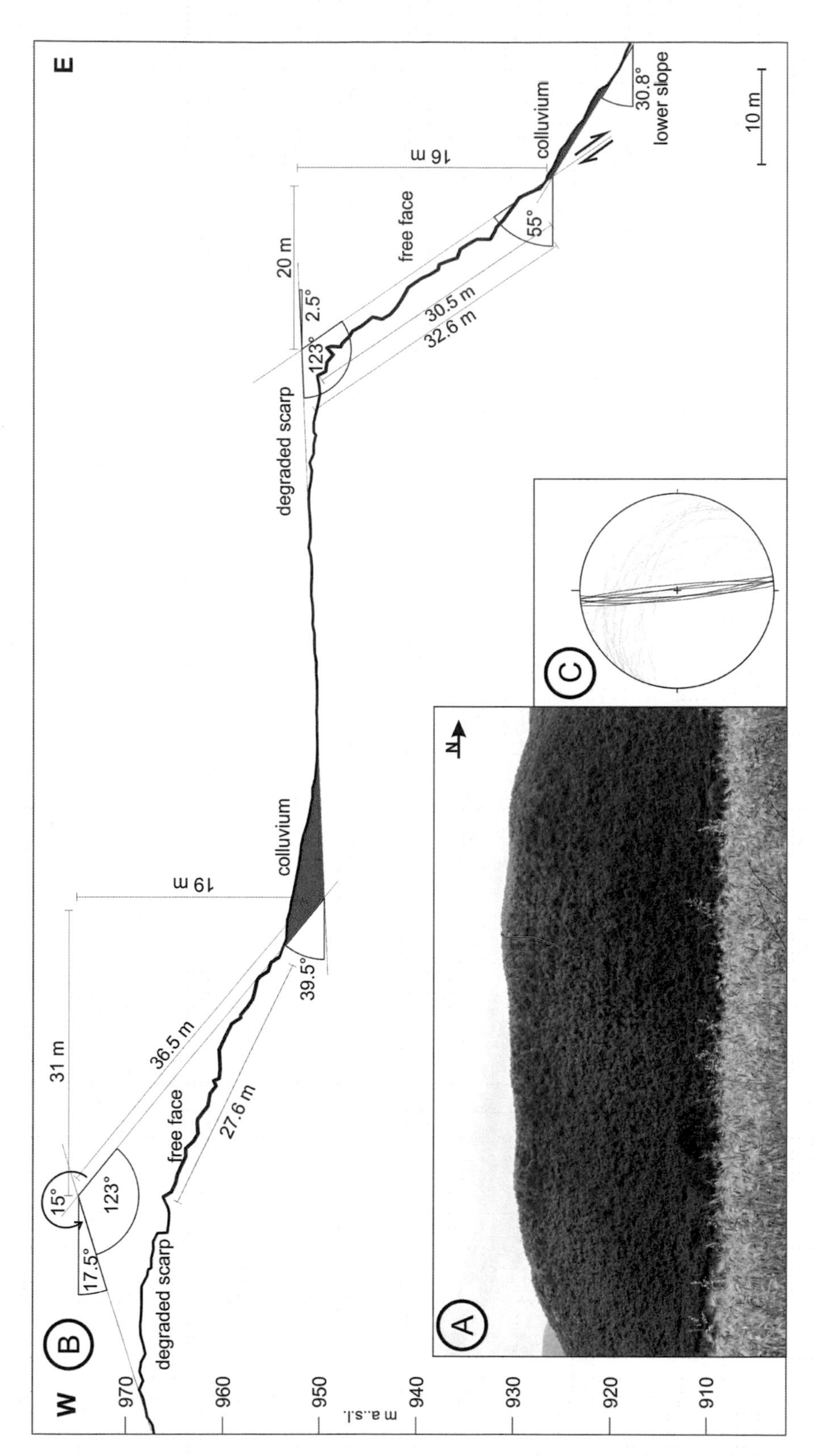

Figure 6.20.: A: Photograph of the Boces scarp. Yellow line gives the approximate position of the profile. B: Complete profile and geometric reconstruction of the Boces fault ($41°02'50''N, 20°37'30''E$; 967 m a.s.l., for location see figure 6.1). No vertical exaggeration. Green lines show constructed angles of upper and lower slope and the fault plane. Red lines indicate the free face measured in the field. The colluvium (marked in yellow) is constructed from field data. C: Stereonet plot of measurements from the fault plane. Light grey plotted planes were measured at the lower, highly jointed part of the scarp, while the planes plotted in black were measured at a well preserved part of the fault plane.

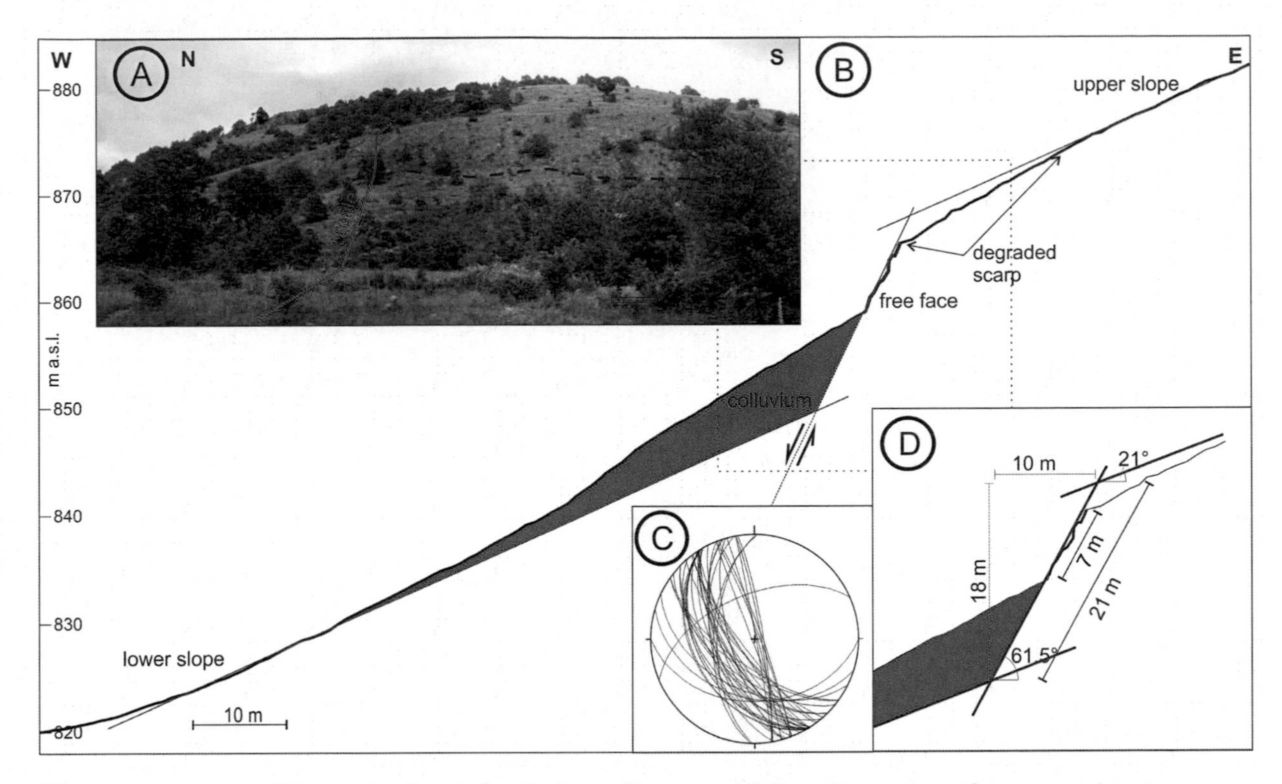

Figure 6.21.: A: Photograph of the Delogozdi scarp. Yellow line gives the approximate position of the profile. Dashed red line gives the position of the free face. B: Complete profile and geometric reconstruction of the Delogozdi fault ($41°15'25''N, 20°43'53''E$; 866 m a.s.l. for location see figure 6.1). Vertical exaggeration is 1.25. Green lines show constructed angles of upper and lower slope and the fault plane. Red lines indicate the free face measured in the field. The colluvium (marked in yellow) is constructed from field data. C: Stereonet plot of measurements from the fault plane. D: Dip angles and offset derived from geometric construction.

the fault scarp can clearly be seen but it is generally smooth and not extensive enough to create the underlying colluvium, which has about three to four times the volume of the degradation. It is most likely that the colluvium, which is a mixture of phyllites, limestone fragments and soil, originates from above the scarp and is deposited here.

6.3. Discussion

The Lake Ohrid Basin, especially the eastern (Galicica Mountain Range) and western flanks (Mokra Mountain Range), are dominated by normal faulting, and the northern and southern parts (the plains of Struga and Pogradec; see Hoffmann et al., 2012) are dominated by clastic input. In general, the north and the south do not seem to be as active although indications for tectonic activity were observed here.

The geology of the basin, which is dominated by two major units (see chapter 3.3) plays a major role in the distribution of the morphotectonic features. In the Mirdita Ophiolite Zone, Triassic conglomerates and Jurassic ophiolites show weak resistance to weathering processes and therefore exhibit less or less pronounced fault scarps. The only exceptions to this are the Upper Cretaceous limestone caps along the southwestern shoreline. In contrast, the Korabi Zone is dominated by Palaeozoic metamorphic rocks and Triassic limestones, which are more resistant to weathering and therefore create the staircase-like morphology that is typical for the basin. The result of the lithological zonation is an inhomogeneous morphological surface expression within the influence of the same stress field. It can also be observed that the NW-SE striking lithological contacts are crosscut by the N-S strike of the basin and its young normal faults. Older structural features like folds, faults, and joints are also cut by these N-S trending normal faults.

The main trend of the fault scarps is N-S, with variations of $\pm$ 20°, and dips ranging between 42° – 85° (along the western shore dips are to the east, and on the eastern shore dips are to the west), hence, forming a graben structure. This observation correlates with the data from the palaeostress analysis (4), where the E-W extensional phase is the youngest. Normal displacement predominates with a minor strike-slip component, which is both, dextral and sinistral (oblique faults) depending on the site. The heights of the fault scarps located on the hillslopes of Galicica and Mokra Mountains are generally between 2 and 45 m (table 6.1), fault lengths vary between 10 and 20 km and consist of several segments. According to Wells and Coppersmith (1994), these faults are expected to create earthquakes of M 6.5 - 7.0 and associated displacements of 0.6 to 2 m. This estimate is also supported by the data of the 1911 earthquake with a magnitude of 6.7 measured in the Ohrid-Korca region (see also chapter 3.2). It therefore can serve as a reference earthquake for the expected magnitudes.

The distance between the western and eastern shore is c. 15 km; the maximum width of the complete basin is c. 24 km and can be regarded as the width of the graben-like fault zone. The lake area covers 358 km^2, which matches the estimated area of 360 to 500 km^2 needed to have the potential for earthquakes of magnitude 6 to 6.5 (Dramis and Blumetti, 2005). The above mentioned rhombus shaped arrangement of normal faults detected by Lindhorst et al. (2012b) is an indication for the initial opening of the basin in a pull apart geometry (see also chapter 4 for comparison). Later E-W extension accounts for the young N-S trending normal faults bounding the graben shoulders. The faults found offshore can be traced onshore (Lindhorst et al., 2012b), so that the stress field was stable since the initial rifting phase affecting the basin. The idea that the active faults become younger towards the center of the basin is supported by data from Lindhorst et al. (2012a,b). But, in their study no evidence for fault activity was detected along the faults located in the deep central part of the basin. The younger faults are observed outside of the central area framed by the two faults called Pescani and Piskupat by Lindhorst et al. (2012a). In contrast, sets of large normal faults were identified by Lindhorst et al. (2012a) along both sides of the lake, which offset Holocene sediments and therefore point to recent activity.

Of 36 measured faults, six were not analysed as their quality was not sufficient. Three of these are located in softrock and did not qualify for the calculation of mean values and slip rates. 19 fault scarps can be defined as planar non-rotational faults (McClay and Ellis, 1987; Wernicke, 1981, 1982) with an upper slope angle that mostly matches the lower slope angle. These faults show mainly a vertical displacement component. This fault type is generally found at symmetrical grabens where fault bound blocks are downthrown.

Another 11 scarps show rotation towards the fault plane, with only four of these being located on the east coast. Rotation takes place around an axis parallel to the strike of the fault (Twiss and Moores, 2007). The blocks can rotate until they reach a very low dip and therefore can accommodate large extension. There are three main mechanisms for the observed tilting of the fault blocks. On a listric fault, a huge gap expands between footwall and hangingwall blocks by applying vertical stress. This gap is compensated by a rollover anticline, which is generally associated with listric normal faulting. Otherwise, sets of antithetic faults in the hangingwall accommodate the strain (Twiss and Moores, 2007).

Another deformation mechanism on listric faults is the development of synthetic fault blocks that act like dominoes. These rotate with fault slip and lean against each other. As they rotate, the dip of the faults becomes shallower and increases horizontal displacement significantly. This allows for a larger amount of extension than non-rotational faults. The angle between bedding and fault planes remains constant which means that faults and fault blocks rotate simultaneously at equal rate no matter how far away from the main fault they are located.

The third mechanism involves a set of parallel listric faults. As they slip along the fault planes separating each other, the fault blocks must deform to align along the shape of the main fault (Wernicke, 1981, 1982). Here, the dip of the bedding increases with distance to the main fault and therefore the dip of fault planes decreases with distance to the main fault.

The amount of scarp rotation in the Ohrid Basin varies between 5° and 11° to horizontal, with some exceptions. The influence of erosion and the true rotation can only be estimated. The mean slope angle is 22° throughout the basin. This angle plus the tilt towards the fault have been added to all rotated faults to determine the shape of the corrected fault angles. The result is that most of the corrected angles range between the high 70°s, or low 80°s. Only Hudenisht, Koritsi Rid, Pestani 2 and Sveti Spas reach angles higher than 90°. This implies that, in most cases, a rotation of around 22° is sufficient to reconstruct the original shape of the faults. Therefore, the domino or planar rotational faults model (McClay and Ellis, 1987; Wernicke, 1981, 1982, see figure 6.19) is a possible deformation mechanism. Rollover anticlines and antithetic faults were not observed onshore, but were both found in offshore seismic data by Lindhorst et al. (2012b). In addition, fault plane dips become shallower towards the basin centre (25° at the Lini Fault after Lindhorst et al., 2012b), which is more representative of the model of parallel listric faults. The fault activity is seen in multichannel-seismic sections with scarps and syntectonic rotation of strata (Lindhorst et al., 2012b). But as the dip of beddings was not measured onshore, neither a definite statement can be given, nor the amount of extension can be calculated. Most likely, the basin shows two modes of listric faulting and therefore exhibits a combination of rotated blocks and sets of parallel listric faults.

There is no direct correlation between either the scarp height or dip of the fault plane with altitude. The mean fault height for the west coast is 38.5 m (measured) and 39.5 m (reconstructed); for the east coast it is 25 m (measured) and 29.3 m (reconstructed); and for the north it is 10.4 m (measured) and 17.6 m (reconstructed). The mean dip angle is 55° for the west coast, 67° for the east coast, and 49° for the north. Also, no direct spatial correlation was observed concerning scarp heights (e.g. generally higher scarps towards the basin centre). Only on a local level can it be observed that those scarps, which are located further away from the centre of the basin are in general the higher ones (e.g.

Boces scarp). This observation is probably related to the small-scale disposition of the basin, so that lithology, joints, the impact of the hydrological system, and to a smaller extent the human influence, control the expression of the scarp profiles.

The eastern scarps have a mean angle of 67° and are therefore about 12° steeper than the the west coast (55°). The mean height of the west coast fault scarps is c. 39 m, which is about 10 m higher than the mean values of the fault scarps along the east coast (29 m). Even when the extraordinary fault scarp of Sveti Arhangel is not taken into consideration for the west coast, the mean value does not drop below 37 m. The scarps to the north have a mean value of c. 17 m.

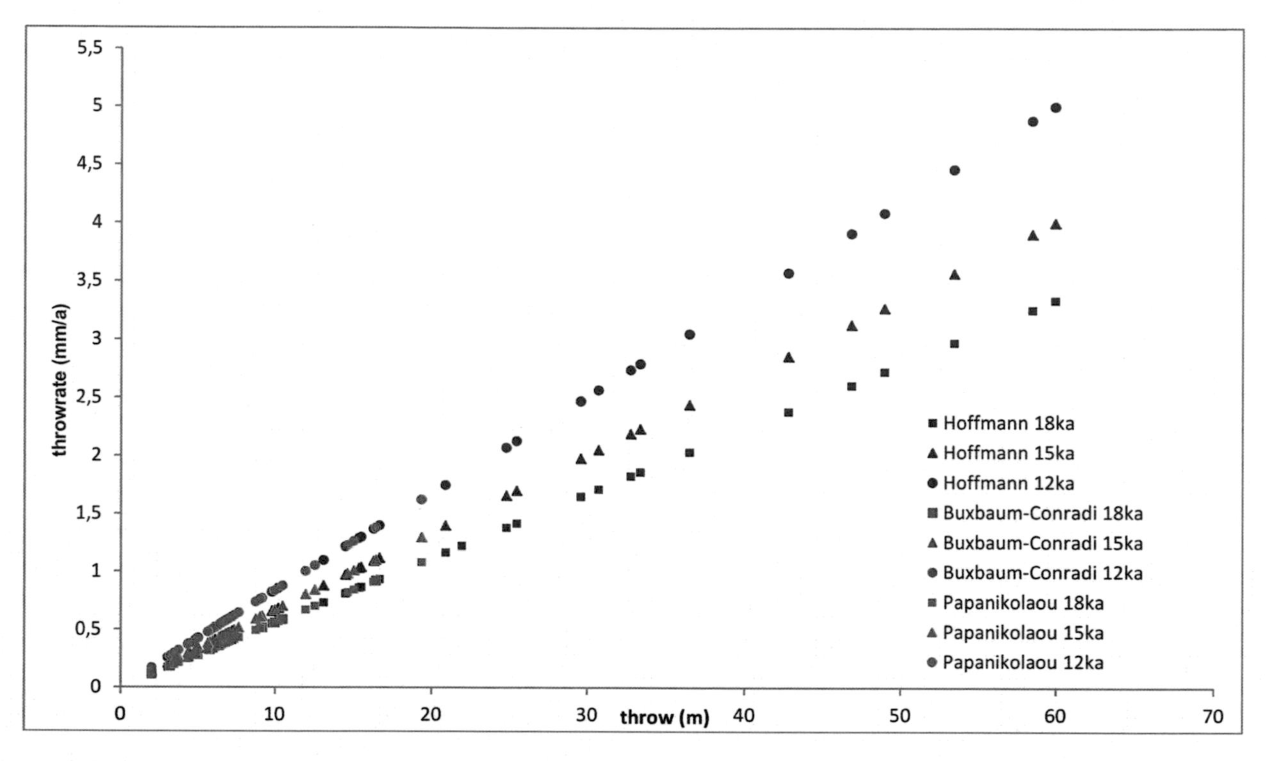

Figure 6.22.: Plots of throw rates against throw with data of various authors. The calculations were done for maximum fault scarp ages of 12ka, 15 ka, and 18 ka.

The Dolno Konjsko outcrop shows that fault scarps with a distinct morphological relief around Lake Ohrid are most likely younger than 11 ka. Because of this assumption and data from other authors (Buxbaum-Conradi, 2010; Papanikolaou et al., 2005; Sadler, 2010, see fig. 6.22;), slip rates were calculated, inferring a postglacial development of the exposed fault scarps for the last 12, 15 and 18 ka. Slip rate values range between 0.42 (0.33, 0.28) mm/a and 4.87 (3.90, 3.25) mm/a for the last 12 (15, 18) ka respectively. These values exceed by far the slip-rates calculated by (e.g. Benedetti et al., 2003) who were working on dated limestone fault scarps in Greece with slip rates of 0.2 mm/a. Considering the seismic activity of the Ohrid-Korca area, this result is not satisfactory. The strong variability of scarp heights (some meters to some 10 meters) in the research area would mean that:

(1) the postulated age of the higher scarps is simply too young, or

(2) the higher the scarp is, the faster the movement on the fault plane must be, unless the beginning of scarp preservation can be pinpointed to a certain time.

Comparing slip data from Papanikolaou et al. (2005) with data from the Ohrid Basin shows that slip rates increase with the amount of slip. Therefore, the simple idea of postglacial development of fault scarps is not sufficient to explain the height of fault scarps; Especially taking into account that there is no correlation between scarp height and position in the basin. The thickness of the colluvium was only constructed from the graphics and evaluated by field observations but not confirmed by drilling. In addition, spatial variations in erosion rates can be a factor that has an influence on the constructed fault geometry. However, these factors can not be responsible for the differences in slip rates. A theory that could cope with the problem is the concept of "stop and go" faults. Here, the older outer faults would slip every time a younger fault evolves and therefore create a higher relief than the younger fault scarp. If considering that the outer fault slips with every seismic event, it is therefore higher the older it is, i.e. the more successive events it has been subject to (see Boces scarp).

As the there is no pattern applicable to the locations of rotated faults, the idea is that not only the faulting mechanism but also gravitational forces play a major role in scarp displacement. In the Hudenisht area in the southeast of the basin (see fig. 6.8A, B), a set of four parallel faults is observed. If a closer look is taken on the river courses (fig. 6.8A, B), it is obvious that the valleys are not straight but show a major bend towards south at the second fault from east. This arrangement can be the result of either a transtensional component or more likely, of mass wasting processes, where the separate blocks are transported towards the basin centre. In this process the blocks are again offset against each other and rotated. Lindhorst et al. (2012a) observed a major underwater slide offshore Hudenisht ("Udenisht Slide Complex" in Lindhorst et al., 2012a) covering an area of c. 27 km^2 and extending approximately 10 km into the central basin. This slide complex is linked to N-S striking normal faults and was possibly triggered by the AD 515, 526 or 527 earthquake (Wagner et al., 2012). Along with other mass wasting deposits seen in seismic data, this shows that the western slope is in general unstable. In the lake sediments this might be linked to high pore pressures, which does not apply onshore. A more sufficient explanation is that the highly fractured rocks, in combination with predefined and easily reactivated fault planes acting as single blocks, are prone to react to seismic events or large offshore mass wasting processes.

Landforms such as a big E-W trending wind-gap are the result of an older faulting mechanism in Late Eocene to Oligocene (see also chapter 4). Below the wind gap and associated with the hanging valley in fig. 6.9, the scarps of Sveti Spas and Lako Signoj are located along the same lineament, which is the front of a halfgraben below the hanging valleys (see figs. 6.9). Parallel to the shoreline extensional gashes or landslide break offs are observed, which indicate mass wasting processes.

The triangular facets found in the lake Ohrid Basin are well distributed and are only in parts highly dissected. But their general form is still obvious. Hanging valleys at the slope of Galicica and Mali I Thate mountains also testify to normal faulting events. But as the valleys are only seasonally water bearing the erosional force for an equilibrium state is low. The morphological expression of seismic induced landforms such as triangular facets is highly dependent on different factors such as lithology, erosion mechanisms, water supply, climate, vegetation cover, anthropogenic modifications, etc. Therefore, all approaches considering tectonomorphological evidence need to be proven with other techniques and/or supported by enough measurements to minimise errors. Furthermore,

Table 6.1.: Table of all scarp profiles. Including position, lithologies and geometric parameters.

Location	coordinates	height [m a.s.l.]	lithology	throw free face [m]	heave free face [m]	offset free face [m]	throw construction [m]	heave construction [m]	offset construction [m]	fault angle	rotation [°]	slope angle upper	slope angle low er	corrected fault angle
Bebien	41°12'35.1"N 20°37'54.5"E	817	limestone	21,96	8,87	23,68	21,96	8,87	28,13	21,74	2,5 ccw	2,5 ccw	23,27	46,24
Boces upper part	41°02'50.7"N 20°37'31.1"E	965	limestone	11,94	24,86	27,58	19,355	30,99	36,54	39,5	17,5 ccw	17,5ccw	2,5ccw	79
Boces low er part	41°02'50.7"N 20°37'31.1"E	965	limestone	23,05	20,095	30,58	25,865	19,935	32,66	55	2,5 ccw	2,5ccw	30,8	79,5
Boces softrock	41°02' 35"N 20°37'17,9"E	956	ophiolith	2,645	9,505	9,87	2,645	9,505	9,87	16,2	4,4 ccw	4,4ccw	30,8	42,6
Delogozdi 1	41°15'29,3"N 20°43'56,2"E	898	limestone	5,54	1,225	5,67	7,21	0,99	7,28	83	x	21	21	
Delogozdi 2	41°15'25,6"N 20°43'53,5"E	882	limestone	6,105	3,61	7,09	18,34	10,075	20,93	61	x	9	25	
Dzepin	41°15'18,1"N 20°44'08,1"E	943	limestone	12,085	14,07	18,55	20,29	14,275	24,81	55	x	21	21	
Elen Kamen	41°08'23,6"N 20°38'51,4"E	767	limestone	25,435	8,28	26,75	32,21	6,345	32,83	78,9	x	21,9	2,3	
Elsani 1	41°01'21,3"N 20°49'19,5"E	1076	limestone	10,565	3,89	11,26	15	4,13	15,56	75	x	5,6	5,6	
Elsani 2 upper part	41°01'19,6"N 20°49'18,0"E	1040	limestone	3,005	0,63	3,07	3,055	0,195	3,06	75,8	x	19,6	19,6	
Elsani 2 low er part	41°01'19,6"N 20°49'18,0"E	1038	limestone	4,825	2,75	5,55	5,385	3,025	6,18	61,4	x	19,6	19,6	
Elsani 3	41°01'15,1"N 20°49'21,0"E	1092	limestone	3,55	4,135	5,45	5,975	3,3	6,83	61,2	4,5 ccw	11,3	29,7	87,7
Elsani 4 total	41°01'11,2"N 20°49'26,1"E	1125	limestone	23,81	19,83	30,99	41,35	12,06	43,07	70,5	x	21,6	32	
Elsani 5	41°02'3,9"N 20°48'43,3"E	856	limestone	120,855	127,325	175,55	120,855	127,325	175,55	42,5	x	16,44	?	
Hudenisht	40°57'08,9"N 20°37'42,8"E	1121	limestone	24,685	14,205	28,48	58,47	9,35	59,21	81	5 ccw	5 ccw	25	108
Hudenisht softrock	40°57'12,6"N 20°38'28,3"E	761	sandstone	26,88	47,125	54,25	29,235	4,675	29,61	29,9	x	10,5	2,9	
Koritsi Rid	40°57'55,9"N 20°48'40,8"E	1425	limestone	18,78	13,055	22,87	29,555	15,655	33,45	62	11 cw	11 cw	38	95

Table 6.1.: (continued)

Location	coordinates	height [m a.s.l.]	lithology	throw free face [m]	heave free face [m]	slip free face [m]	throw construction [m]	heave construction [m]	slip construction [m]	fault angle	rotation [°]	slope angle upper	slope angle low er	corrected fault angle
Lako Signoj upper part	40°58'11"N 20°48'7,1"E	1084	limestone	11,04	2,375	11,29	14,48	1,44	14,55	85,5	x	25,5	25,5	
Lako Signoj low er part	40°58'11"N 20°48'6,1"E	1076	limestone	10,66	9,545	14,31	12,505	9,02	15,42	54	x	25,5	25,5	
Lini 1	41°04'10,4"N 20°37'31,6"E	908	limestone	6,96	5,715	9,01	9,395	2,9	9,83	81,7	x	20,4	20,4	
Lini 2	41°04'11.0"N 20°37'31,1"E	897	limestone	17,57	45,625	48,89	17,57	45,625	48,89	21,2	x	19,94	19,94	
Lini 3	41°03'57,7"N 20°37'57,4"E	759	limestone	8,605	9,655	12,93	6,395	2,445	6,85	69	x	21,47	29	
Memelisht	40°55'10,1"N 20°38'14,9"E	947	limestone	31,775	30,335	43,93	27,69	13,35	30,74	64	x	17,5	22	
Memelisht softrock	40°56'31.30"N 20°37'27.60"E	1040	harzburgit	13,055	51,86	53,48	13,055	51,86	53,48	15,4	4,3 ccw	4,3ccw	26,9	
Pestani 1	41°00'44,7"N 20°48'47,3"E	822	limestone	2,84	0,885	2,97	4,93	1,155	5,06	77	x	8	8	
Pestani 2	41°00'47,4"N 20°48'48,2"E	799	limestone	5,22	3,145	6,09	5,475	1,915	5,80	71	1,5 cw	1,5 cw	28	94,5
Piskupat	41°02'16,2"N 20°37'33,9"E	916	ophiolite/ limestone	25,395	34	42,44	31,82	40,33	51,37	39	22 ccw	22 ccw	11	83
Sveti Arhangel 1	41°06'27,4"N 20°37'56,5"E	840	limestone	42,975	0,75	42,98	46,86	0,835	46,87	89	x	27,2	27,2	
Sveti Arhangel 2	41°06'28,5"N 20°37'55,4"E	845	limestone	50,37	7,965	51,00	57,825	9,09	58,54	81	x	28,12	28,1	
Sveti Stefan	41°04'22,5"N 20°48'19,1"E	1031	limestone	7,375	3,69	8,25	10,2	0,86	10,24	85	x	2,5	27	
Sveti Spas	40°59'30,1"N 20°48'43,8"E	1070	limestone	3,835	1	3,96	4,375	0,965	4,48	77,3	1,5 cw	1,52	23,25	100,8
Velestovo 1	41°5'14,5"N 20°49'17,2"E	985	limestone	11,445	11,02	15,89	12,5	11,135	16,74	65	x	0,98	24,3	
Velestovo 2	41°5'6,3"N 20°49'23"E	1004	limestone	41,945	39,345	57,51	47,825	36,21	59,99	87,3	x	6,23	28,88	
Velestovo 3	41°04'16,5"N 20°49'08,0"E	1250	limestone	29,165	21,775	36,40	34,22	25,74	42,82	53	11,5 cw	11,5 cw	26,8	86,5

joints, foliation or bedding planes parallel to a fault, even in inactive tectonic settings, can create straight mountainfront-piedmont junctions and well defined triangular facets, whereas the influence of big streams with high erosional power is responsible for a quicker dissection of the triangular facets. A summary of major neotectonic features is presented in figure 6.23.

According to the results of fault scarp profiling the western flanks dipping with 55° shallower than the east coast (67°). Which is in contrast with data from Lindhorst et al. (2012b), which show a steep dipping basement towards the west and a gentler dipping slope towards the east. Also the height of the fault scarps is 39.5 m and therefore higher than fault scarps along the eastern flanks where there is a mean hight of 29.3 m. Therefore, the western flank of the basin seems to accommodate for most of the extension. Data of Lindhorst et al. (2012b) depicted large sediment cover on the western slopes but also found evidence for active faulting especially along the Lini Fault that exhibits a shallow angle at 25°. While the east coast is highly segmented with tilted blocks of basement, the west coast is dominated by mass wasting processes most likely triggered by seismic events or by the removal of material below so that the slope becomes unstable. Possibly these mass wasting processes are responsible for the shallower angles and higher fault scarps, by causing slip along the fault planes without the trigger of a seismic event.

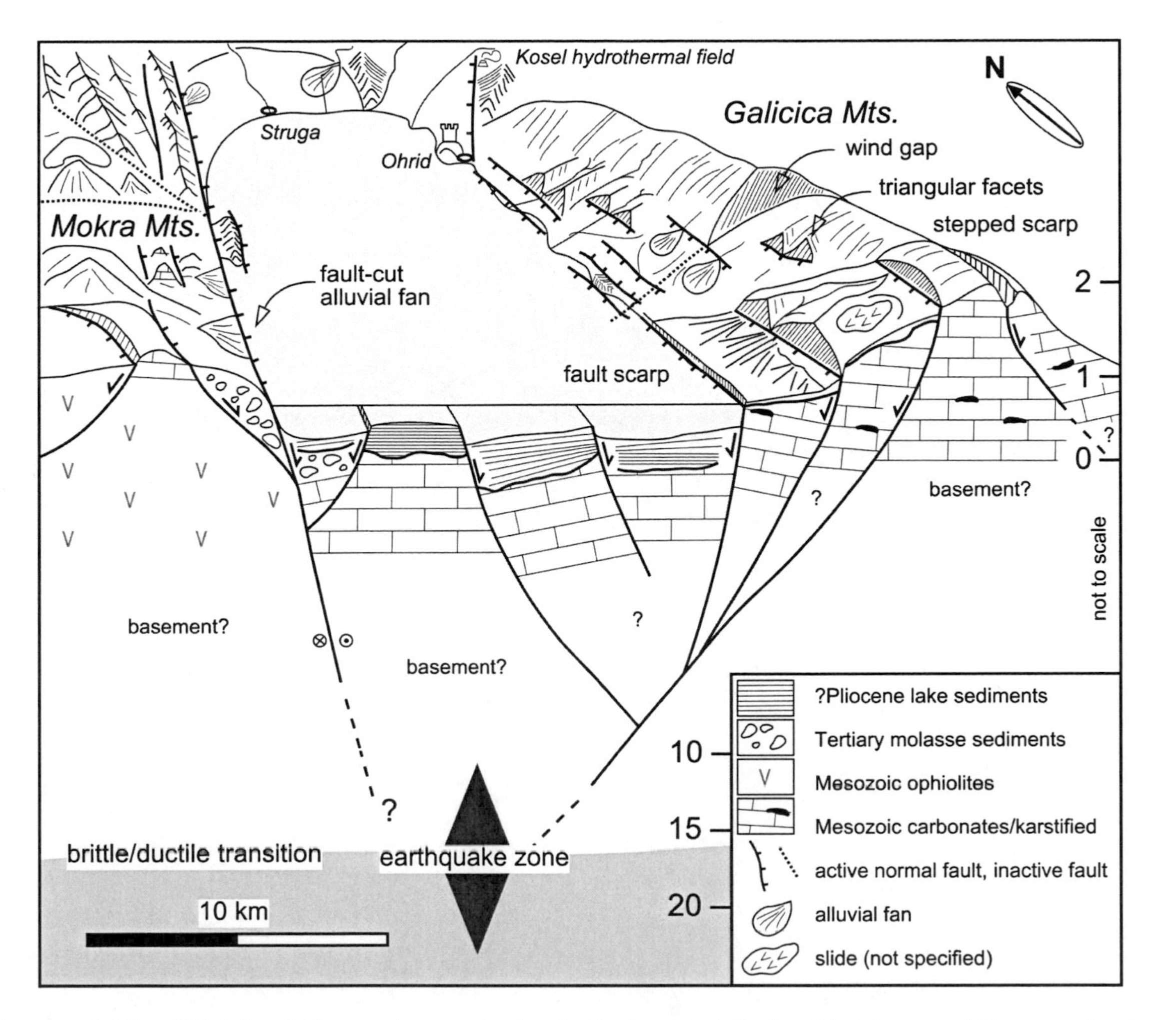

Figure 6.23.: Sketch of the major neotectonic features at Lake Ohrid resembling a seismic landscape (basal section from an example of the Italian Appenines, modified after Dramis and Blumetti, 2005). Note the Kosel hydrothermal field along a major fault in the north. The earthquake zone confines the focal depths of instrumentally recorded earthquakes in the area, which is between 12 and 25 km depth. The western basin-bounding fault has a dextral strike-slip component. Modified after Reicherter et al. (2011).

7. Synthesis

This thesis comprised detailed studies into the tectonic evolution and seismicity of the Lake Ohrid Basin. This was carried out through investigations of the palaeostress regime, the sedimentary mechanisms and of the geomorphological evidence, which provided information on fault geometries, slip rates, landscape evolution, the grade of activity of faults, etc. The data from these studies were used to determine how the landscape has evolved with respect to seismicity. Here, a summary of the conclusions that have been made individually in the preceding chapters and their relevance for the defined objectives is presented.

Determination of the palaeostress fields which controlled the evolution of the lake

The evolution of the palaeostress field over time was revealed by an intense palaeostress analysis. Five successive periods of basin development were derived from the data and embedded into the timing of major shifts in the geodynamic setting.

- Orogenic Phase
  - Cretaceous-Paleogene: NE-SW compression
  - Late Eocene-Oligocene: NE-SW extension
  - Oligocene-Miocene: NE-SW shortening and strike-slip movement accompanied by oblique movements
- Mid-Miocene: Transtensional phase with NW-SE extension and strike-slip movement. The opening of the basin nucleated along a mirrored "s"-shaped releasing bend, in a dextral strike-slip mode. This led to the development of an extensional duplex and the evolution of a pull apart basin. The dextral shear is also indicated by a large E-W trending wind gap, which finds its counterpart west of the Lini Peninsula.
- Late Miocene to recent: Change of the stress field to pure E-W extension.

The various changes of stress fields have left their imprints as highly fractured rocks with reactivations and overprinting of old joint and fault systems. This is also reflected in the differences in deformation at different scales. While on a large scale, the major N-S oriented faults give the basin its straight shorelines and its N-S stretched appearance, on a smaller scale the coastline is highly segmented. In contrast to the general trend, the smaller faults strike NW-SE or NE-SW which results in the zigzag line of the coast. The strong segmentation can also be seen in the lake, where several smaller faults take up the stress.

A maximum extension rate of c. 10.5 mm/a was calculated assuming the proposed minimum age of 1.9 Ma (Lindhorst et al., 2012b) for the basin and taking an average basin width of 20 km. Data from other basin and range provinces (Eddington et al., 1987; Thatcher et al., 1999) suggest overall extension rates between 8-12 mm/a with locally increasing or decreasing values. The values for the Lake Ohrid Basin are therefore considered reasonable.

Information on fault orientation, fault geometry, spatial distribution of fault scarps across the basin, the influence of lithology on faulting mechanisms, and slip rates.

The fault orientation is mainly N-S with variations of ± 20° concerning the youngest deformational phase. There are two types of fault geometry. Faults with a staircase-like appearance, and faults which have been rotated resulting in back tilting. The tilted faults were rotated back to their original position for a better comparison with non-tilted faults. The analysis of fault scarps revealed the average fault geometry. The upper slope dips at 22°, beneath this the scarp's free face dip angle ranges between 42° and 85°; mean dips are 55° for the west coast, 67° for the east coast, and 49° for the north. The mean constructed fault height is 39.5 m for the west coast; 29.3 m for the east coast; and 17.6 m for the north. Below the fault plane colluvium is present. The colluvium has a variable distribution depending on the transport mechanisms, the steepness of the lower slope and geomorphological particularities. The lower slope angle normally matches the upper slope angle and is therefore also at 22°.

The spatial distribution of faults measured in the presented study is mainly concentrated along the eastern and western coasts, and a few examples at the northern end of the basin. The spatial occurrence of fault scarps is highly connected to the influence of lithology, which is dominated by the two major units: the Mirdita Ophiolite Zone with weak resistance to weathering processes and less pronounced fault scarps; and the Korabi Zone, where the lithologies create the typical staircase-like morphology. This lithological zonation creates an inhomogeneous fault distribution with high variations in morphological expression for fault scarps produced by the same stress field.

Slip rate values on average range between 0.28 mm/a and 3.25 mm/a for the last 18 ka with higher values calculated for 15 ka and 12 ka. These are very high values compared with other authors (Benedetti et al., 2003, e.g.) who postulate a slip rate of 0.2 mm/a under comparable conditions. This leads in turn to the concept of "stop and go" faults, where a fault would slip every time a younger fault evolves and by doing this it generates a higher slip rate. The observation that those scarps, which are located further away from the centre of the basin are in general the higher ones can only be made on a local level. Thus, lithology, erosion mechanisms, climate, vegetation cover, human modifications, etc. control the expression of the fault scarps and other morphological features (e.g. triangular facets). Also, two modes of listric faulting were observed in the basin creating a mixture of faulting along rotated blocks and sets of parallel listric faults. In general, no pattern could be assigned according to the spatial distribution, geometry or faulting mode of fault scarps. Therefore, also gravitational forces must play a major role in scarp displacement.

Role of alluvial fans and delta formation on landscape evolution around Lake Ohrid.

There are several systems of alluvial fans and delta complexes around Lake Ohrid. The northern plain around Struga hosts large, but very shallow, alluvial fans, which are today mainly inactive. Other alluvial fans were detected along the eastern coast but their lobe shaped structure is cut by normal faults.

River deltas were investigated at the Cerava and the Daljan River in the northern and the southern part of Lake Ohrid respectively. Their meandering fluvial systems have been active at least during Holocene and seem to have been stable over a long period. The areas have been drained in the last decades for agricultural use, which partly led to depletion of the streams. The landscape evolution is, therefore, driven by two morphological systems: the plains to the north and south dominated by sedimentary input and climatically driven alluvial fans and deltas, and the eastern and western hillslopes with a strong influence from active normal faulting.

Identification of areas experiencing tectonic deformation - active / less active areas throughout the basin.

The recent deformation regime has affected the whole basin and is seen in various morphological features all over the area creating a seismic landscape. Not only normal faulting, but also other mechanisms connected to highly fractured rocks lead to a general unstable western slope and play a major role in the basin evolution. The shape of fault scarps, triangular facets and other features connected to the seismic activity is strongly influenced by external factors such as the lithology. Therefore, all approaches to link seismic activity with the tectonomorphological expression and to gain a classification of areas with more or less activity is very limited. What can be seen is a difference of activity between the east/west coast in contrast to the northern and southern areas. The reason is probably that most of the stress is accommodated by N-S trending faults. Thus, the Galicica, the Mali i Thate and the Mokra Mountain ranges are dominated by normal faulting, and the plains of Struga and Pogradec are not as active; the plains are dominated by clastic input although indications for tectonic activity were observed here. The east coast is highly segmented with tilted blocks of basement rock, the west coast is dominated by mass wasting processes. In general there were no large areas onshore in the basin which can be referred to as inactive.

Grade of activity in the Lake Ohrid Basin.

The fault lengths, which vary between 10 and 20 km, determine the grade of activity in the basin. These large faults are considered of being capable of producing earthquakes with magnitudes between 6.0 and 7.0 and associated displacements of 0.6 to 2 m per event. The historical seismic record shows severe earthquakes occur in the region, which support this estimation (e.g. the magnitude 6.7 earthquake in 1911). Frequent shallow and low magnitude earthquakes are recorded by seismic stations (e.g. June 8th 2012, M_b 4.4; 41.24°N, 20.90°E; focal depth ~10km, normal faulting; EMSC, 2013; NEIC, 2013). Seismic features and dating of a dragged palaeosol testify to an active seismic landscape in Holocene times. The area is therefore under consideration for events in a range that can cause serious damage to infrastructure. The case of the Debar earthquake 2009-09-06 with a magnitude of 5.6 shows that even earthquakes this size can cause serious damage to man-made structure partly due to poor building standards. At Lake Ohrid especially, the zones between sloped and flat areas are popular areas for housing. These areas are highly vulnerable to ground shaking and by secondary seismic effects like liquefaction, seeps, dewatering structures, rock falls and landslides and others (Reicherter et al., 2009).

Determine if identified tectonic structures are capable of causing major speciation events and highlight possible triggering mechanisms.

The small-scale internal morphology creates a large variety of habitats, which can be occupied by various species. Vast flat plains contrast the steep slopes with blocks creating small basins and niches. The step like morphology creates vertically separated and varying environmental conditions (by means of light, temperature, etc.).

Moreover, events with magnitudes larger than 6.0 can cause a large input of material into the lake (e.g. by landslides, rockfalls, etc.). These events then change the living conditions over a longer period of time e.g. by the input of small particles. Suspended particles stay within the water column with long retention times and therefore change the input of light, availability of nutritional sources, the pH-value, etc. This can add extra stress to a population and cause a speciation event. Available earthquake data only date back to historical times and data from short cores and OSL dating suggest events around 4,500 a cal. BP and at 11,000 a BP respectively. Older records obtained from e.g. the SCOPSCO drilling campaign possibly contain more events and can then be correlated with molecular clock analyses to find concordant events. This might give an idea about tectonic-related speciation in Lake Ohrid. Possible triggering mechanisms driving speciation can be mass wasting processes, change of the location of rivers and streams, earthquakes and secondary seismic effects, morphological segmentation connected to tectonic activity associated with rapid changes of habitats, and the high variability of habitats linked to these processes.

This study has illustrated that by combining different fieldwork techniques including geological and geomorphological mapping, geophysical reconnaissance, drilling/coring and remote sensing with advanced analytical techniques, our knowledge of the seismicity and tectonic evolution of an area can be significantly advanced. To get a better idea on slip rates and offset per event trenching across active faults would be useful. This had to be postponed in the present study due to a lack of appropriate trenching sites. The rough ground conditions, with a high water table and sediments containing large amounts of angular gravel were not considered to be able to preserve the desired sedimentary structures. Another drawback was the inaccessibility of sites, due to housing or other infrastructure. Future steps of work besides trenching studies would be mapping and dating of secondary environmental earthquake effects in order to generate a susceptibility and hazard map for the Lake Ohrid Basin.

Acknowledgements

I left this acknowledgements section till the very last minute, so I am in a rush and because I am forgetful at the best of times I want to say sorry to those who feel to have been forgotten here. This work would not have been possible without my supervisor Prof. Dr. Klaus Reicherter as he had the confidence in me to run the Lake Ohrid Project and did a great job to teach me all this fieldwork stuff. He provided me with hot discussions and the inspiration to think out of the box. Alessandro Michetti as my second reviewer should be mentioned as he accompanied my career since we met in Israel the first time and provided me with lots of information and discussions.

Great support came from the Hydrobiological Institute of Ohrid providing us with housing, permissions, equipment, and a good laugh. Especially Zoran Spirkovski and Dusica Ilic-Boeva became friends over the years. Special thanks go to Captain Zoran Brdarovski for all kinds of technical support. The support and backup of the great team of the Neotectonics and Natural Hazards Group mean a lot to me. Especially the experiences on countless field trips for good and for bad with Dr. Christoph Grützner, Thomas Wiatr and Andreas Rudersdorf will always have a warm place in my memories. Jack Mason was a great help by improving my English skills and showed a lot of patience while correcting the same mistakes continually. Thanks also to those who taught me that a good dirty day in the field should be accomplished with a proper sundowner somewhere at the beach.

I would also like to thank all the students that have contributed to the work and gave their manpower for drilling, measuring and climbing the mountains at Lake Ohrid: Katharina Wohlfahrt, Dorothee Uerschels, Nina Engels, Eva Hölzer, Tim Krüger, Sandra Fuhrmann, Max Oberröhrmann, Ariane Liermann, Rebecca Peters, Melanie Walter, Marc Sadler, Christian Buxbaum-Conradi, Miroslav Malicevic and Christian Diebel.

This work is based on the Lake Ohrid Project, which was financially supported by the Deutsche Forschungsgemeinschaft (project "Sedimentary and Neotectonic History of Lake Ohrid (FYROM/Albania)" Re 1361/10).

Last but not least I want to thank my family, who laid in every aspect the foundation to follow my way and from the very beginning supported my affinity to rocks in particular, my close friends and my partner Marcell who had to endure all ups and downs over the years.

References

Albrecht, C., Wilke, T. (2009): Ancient Lake Ohrid: biodiversity and evolution, Hydrobiologia, 615, 103–140.

Albrecht, C., Trajanovski, S., Kuhn, K., Streit, B., Wilke, T. (2006): Rapid evolution of an ancient lake species flock: Freshwater limpets (Gastropoda: Ancylidae) in the Balkan Lake Ohrid, Organisms Diversity and Evolution, 6, 294–307.

Aliaj, S., Sulstarova, E., Muco, B., Kociu, S. (1995): Seismotectonic Map of Albania, Academy of Science; Seismological Institute Tirana.

Aliaj, S., Adams, J., Halchuk, S., Sulstarova, E., Peci, V., Muco, B. (2004): Probabilistic Seismic Hazard Maps for Albania, in: 13th World Conference on Earthquake Engineering. Vancouver, B.C., Canada, Paper No. 2469.

Ambraseys, N. (2009): Earthquakes in the Mediterranean and Middle East: a multidisciplinary study of seismicity up to 1900, Cambridge University Press, New York.

Ambraseys, N., Jackson, J. (1990): Seismicity and associated strain of Central Greece between 1890 and 1988, Geophysical Journal International, 101, 663–709.

Anderson, E. (1942): The Dynamics of Faulting, Olivier and Boyd, Edinburgh.

Anderson, H., Jackson, J. (1987): Active tectonics of the Adriatic Region, Geophysical Journal of the Royal Astronomical Society, 91, 937–983.

Andree, R. (1886): Die unteren Donauländer zur Römerzeit, in: Professor G. Droysens Allgemeiner Historischer Handatlas, Plate 17, www.pelagon.de.

Angelier, J. (1990): Inversion of field data in fault tectonics to obtain the regional stress; Part 3, A new rapid direct inversion method by analytical means., Geophysical Journal International, 103(2), 363–376.

Angelier, J. (1994): Fault slip analysis and paleostress reconstruction, in: Continental Deformation, edited by Hancock, P. L., chap. 4, pp. 53–100, Pergamon Press Oxford University Press.

Armijo, D., Meyer, B., Hubert, A., Barka, A. (1999): Westward propagation of the North Anatolian fault into the northern Aegean: Timing and kinematics, Geology, 27, 267–270.

Armijo, R., Lyon-Caen, H., Papanastassiou, D. (1991): A possible normal-fault rupture for the 464 BC Sparta earthquake, Nature, 351, 137–139.

Armijo, R., Lyon-Caen, H., Papanastassiou, P. (1992): East-West extension and Holocene fault scarps in the Hellenic arc, Geology, 20, 491–494.

Arsovsky, M., Hadžievsky, D. (1970): Correlation between neotectonics and the seismicity of Macedonia, Tectonophysics, 9, 129–142.

Atwater, R. (1927): Procopius of Ceasarea: The Secret History, P. Covici, Chicago, 1927 and Covici Friede, New York, reprinted, University of Michigan Press, Ann Arbor, MI, 1961,with indication that copyright had expired on the text of the translation,.

Bébien, J., Shallo, M., Manika, K., Gega, D. (1998): The Shebenik massif (Albania), a link between MOR- and SSZ-type ophiolites?, Ofioliti, 23, 7–15.

Benedetti, L., Finkel, R., Papanastassiou, D., King, G., Armijo, R., Ryerson, F., Farber, D., Flerit, F. (2002): Postglacial slip history of the Sparta Fault (Greece) determined by 36Cl cosmogenic dating: evidence for non-periodic earthquakes, Geophysical Research Letters, 29, 87/1–87/4.

Benedetti, L., Finkel, R., King, G., Armijo, R., Papanastassiou, D., Ryerson, F., Flerit, F., Farber, D., Stavrakakis, G. (2003): Motion on the Kaparelli fault (Greece) prior to the 1981 earthquake sequence determined from 36Cl cosmogenic dating, Terra Nova, 15, 118–124.

Bott, M. (1959): The mechanics of oblique slip faulting, Geological Magazine, 96(2), 109–117.

Bull, W. (2007): Tectonic Geomorphology of Mountains. A New Approach to Paleoseismology., Blackwell Publishing.

Burbank, D., Anderson, R. (2001): Tectonic geomorphology, Oxford (Blackwell).

Burchfiel, B. C., King, R. W., Todosov, A., Kotzev, V., Dumurdzanov, N., Serafimovski, T., Nurce, B. (2006): GPS results for Macedonia and its importance for the tectonics of the Southern Balkan extensional regime, Tectonophysics, 413, 239–248.

Burchfiel, B. C., King, R. W., Nakov, R., Tzankov, T., Dumurdzanov, N., Serafimovski, T., Todosov, A., Nurce, B. (2008): Patterns of Cenozoic Extensional Tectonism in the South Balkan Extensional System, NATO Science Series IV Earth and Environmental Sciences, 81, 3–18.

Burton, P., Xu, Y., Qin, C., Tselentis, G., Sokos, E. (2004): A catalogue of seismicity in Greece and the adjacent areas for the twentieth century, Tectonophysics, 390, 117–127.

Buxbaum-Conradi, C. (2010): Morphotektonische Aufnahme von Störungsstufen am Ohridsee (FYROM/Albania), bachelor's Thesis, RWTH Aachen University (2010).

Byerlee, J. (1968): Brittle-ductile transition in rocks, Journal of Geophysical Research, 73(B14), 4741–4750.

Caporali, A., Aichhorn, C., Barlik, M., Becker, M., Fejes, I., Gerhatova, L., Ghitau, D., Grenerczy, G., Hefty, U., Krauss, S., Medak, D., Milev, G., Mojzes, M., Mulic, M., Nardo, A., Pesec, P., Rus, T., Simek, J., Sledzinski, L., Solaric, M., Stangl, G., Stopar, B., Vespe,

F., Virag, G. (2009): Surface kinematics in the Alpine-Carpathian-Dinaric and Balkan region inferred from a new multi-network GPS combination solution, Tectonophysics, 474, 295–321.

Caputo, R., Monaco, R., Tortorici, L. (2006): Multiseismic cycle deformation rates from Holocene normal fault scarps on Crete (Greece), Terra Nova, 18, 181–190.

CGIAR (2013): Consortium for Spatial Information, Tech. rep., `http://srtm.csi.cgiar.org/`, last access 2013-09-09.

Da Silva, A. C., Potma, K., Weissenberger, J. A. W., Whalen, M. T., Humblet, M., Mabille, C., Boulvain, F. (2009): Magnetic susceptibility evolution and sedimentary environments on carbonate platform sediments and atolls, comparison of the Frasnian from Belgium and Alberts, Canada., Sedimentary Geology, 214, 3–18.

Danzeglocke, U., Jris, O., Weninger, B. (2013): CalPal Radiocarbon Calibration Online, Tech. rep., `http://www.calpal-online.de`, last access 2013-07-09.

Dilek, Y. (2006): Collision tectonics of the Mediterranean region: Causes and consequences, in: Dilek, Y., and Pavlides, S., eds., Postcollisional tectonics and magmatism in the Mediterranean region and Asia: Geological Society of America Special Paper, 409, 1–12.

DLR (2013): TerraSAR-X images, Tech. rep., Deutsches Zentrum für Luft- und Raumfahrt, `www.dlr.de`, last access 2013-01-10.

Doblas, M. (1998): Slickenside kinematic indicators, Tectonophysics, 295, 187–197.

Dramis, F., Blumetti, A. M. (2005): Some considerations concerning seismic geomorphology and paleoseismology, Tectonophysics, 408 (1-4), 177–191, doi:10.1016/j.tecto.2005.05.032.

Duggen, S., Hoernle, K., van den Bogaard, P., Rüpke, L., Morgan, J. (2003): Deep roots of the Messinian salinity crisis, Nature, 422, 602–606.

Duller, G. (2008): Single-grain optical dating of Quaternary sediments: why aliquot size matters in luminescence dating, Boreas, 37-4, 589–612.

Dumurdzanov, N., Ivanovski, T. (1977): Geological Map of the Socialist Federal Republic of Yugoslavia, Sheet Ohrid, Geological Survey Belgrad.

Dumurdzanov, N., Serafimovski, T., Burchfiel, B. C. (2004): Evolution of the Neogene-Pleistocene Basins of Macedonia, Geological Society of America Digital Map and Chart Series 1, pp. 1–20.

Dumurdzanov, N., Serafimovski, T., Burchfiel, B. C. (2005): Cenozoic tectonics of Macedonia and its relation to the South Balkan extensional regime, Geosphere, 1, 1–22.

Eddington, P., R.B., S., Renggli, C. (1987): Kinematics of Basin and Range intraplate extension, Geological Society, London, Special Publications, 28, 371–392, doi:10.1144/GSL.SP.1987.028.01.23.

Ellwood, B. B., Crick, R. E., El Hassani, A., Benoist, S. L., Young, R. H. (2000): Magnetosusceptibility event and cyclostratigraphy method applied to marine rocks: detrital input versus carbonate productivity, Geology, 28, 1135–1138.

EMSC (2013): earthquake database, Tech. rep., Centre Sismologique Euro-Méditerranéen - European-Mediterranean Seismological Centre, `www.emsc-csem.org`, `http://www.emsc-csem.org`, last access: 2013-10-05.

Fuhrmann, S. (2009): Felskartierung entlang der Steilküste des Ohridsees vom Gorica Hügel bis zum Autocamp Gradiste & Scarpanalyse mit Hilfe des terrestrischen Laserscanners, Master's Thesis, RWTH Aachen University.

Gawlick, H. J., Frisch, W., Hoxha, L., Dumitrica, P., Krystyn, L., Lein, R., Missoni, S., Schlagintweit, F. (2008): Mirdita Zone ophiolites and associated sediments in Albania reveal Neothetys Ocean origin, International Journal of Earth Science, 97, 865–881.

Goldman, C., Elser, J., Richards, R., Reuter, J., Priscu, J., Levin, A. (1996): Thermal stratification, nutrient dynamics, and phytoplankton productivity during the onset of spring phytoplankton growth in Lake Baikal, Russia, Hydrobiologia, 331, 9–24.

Goldsworthy, M., Jackson, J., Haines, J. (2002): The continuity of active fault systems in Greece, Geophysical Journal International, 148, 596–618.

Gorthner, A. (1994): What is an ancient lake?, in: Speciation in Ancient Lakes, edited by Martens, K., Goddeeris, B., Coulter, G., vol. 44, pp. 97–100, Archiv für Hydrobiologie.

Grad, M., Tiira, T., Group, E. W. (2009): The Moho depth map of the Euopean Plate, Geophysical Journal International, 176, 279–292.

GSHAP (2013): Global Seismic Hazard Assessment Program, Tech. rep., www.seismo.ethz.ch/gshap,, `www.seismo.ethz.ch/gshap`, last access: 2013-09-20.

Hauffe, T., Albrecht, C., Schreiber, K., Birkhofer, K., Trajanovski, S., Wilke, T. (2011): Spatially explicit analysis of gastropod biodiversity in ancient Lake Ohrid., Biogeosciences, 8, 175–188.

Hoeck, V., Koller, F., Meisel, T., Onuzi, K., Kneringer, E. (2002): The Jurassic South Albanian ophiolites: MOR- vs. SSZ-type ophiolites, Lithos, 65, 143–164.

Hoffmann, N., Reicherter, K., Fernández-Steeger, T., Grützner, C. (2010): Evolution of ancient Lake Ohrid: a tectonic perspective, Biogeosciences, 7, 3377–3386.

Hoffmann, N., Reicherter, K., Grützner, C., Hürtgen, J., Rudersdorf, A., Viehberg, F., Wessels, M. (2012): Quaternary coastline evolution of Lake Ohrid (Macedonia/Albania), Central European Journal of Geosciences, 4(1), 94–110.

Hubbert, M. (1951): Mechanical basis for certain familiar geological structures, Geological Society American Bulletin, 62(4), 355–372.

Iben Brahim, A. (2005): Assessment of seismic risk maps and evaluation of seismic vulnerability of historical building heritage in the Mediterranean area, Tech. rep., Report on WP3 of the PROHITECH Programme, FP62002-INCO-MPC-1; Rabat, www.prohitech.com, last access: 2012-02-06.

Jaeger, J., Cook, N. (1979): Fundamentals of rock mechanics, Chapman and Hall, London.

Jordanoski, M., Lokoska, L., Veljanoska-Sarafiloska, E. (2010): The River Sateska And Consequences Of Its Divergion To Lake Ohrid, Balwois Conference 2010 Proceedings, pp. 1–7.

Jozja, N., Neziraj, A. (1998): Geological heritage conservation of Ohrid Lake, Geologica Balcanica, 28, 91–95.

Jurine, L. (1820): Histoire des monocles qui se trouvent aux environs de Genve, Genve et Paris, I-XVI, 1–260.

Kilias, A., Tranos, M., Mountrakis, D., Shallo, M., Marto, A., Turku, I. (2001): Geometry and kinematics of deformation in the Albanian orogenoc belt during Tertiary, Journal of Geodynamics, 31, 169–187.

Kilias, A., Frisch, W., Avgerinas, A., Dunkl, I., Falalakis, G., Gawlick, H. (2010): Alpine architecture and kinematics of deformation of the northern Pelagonian nappe pile in the Hellenides, Austrian Journal of Earth Sciences, 103(1), 4–28.

Klie, W. (1939a): Studien über Ostracoden aus dem Ohridsee: I. Candocyprinae., Archiv für Hydrobiologie, 35, 28–45.

Klie, W. (1939b): Studien über Ostracoden aus dem Ohridsee: II. Limnocytherinae und Cytherinae., Archiv für Hydrobiologie, 35, 631–646.

Klie, W. (1942): Studien über Ostracoden aus dem Ohridsee: III. Erster Nachtrag., Archiv für Hydrobiologie, 38, 254–259.

Kocks, H., Melcher, F., Meisel, T., Burgath, K. P. (2007): Diverse contributing sources to chromitite petrogenesis in the Shebenik Ophiolitic Complex, Albania: evidence from new PGE and Osisotope data, Miner. Petrol., 91, 139–170.

Kostoski, G., Albrecht, C., Trajanovski, S., Wilke, T. (2011): A freshwater biodiversity hotspot under pressure - assessing threats and identifying conservation needs for ancient Lake Ohrid, Biogeosciences, 7, 3999–4015.

Kotzev, V., King, R. W., Burchfiel, B., Todosov, A., Nurce, B., Nakov, R. (2008): Crustal Motion and Strain Accumulation in the South Balkan Region Inferred from GPS, NATO Science Series IV Earth and Environmental Sciences, 81, 19–43.

Lempriere, D. D. (1838): Bibliotheca Classica, W. E. Dean, New York,.

Leng, M., Baneschi, I., Zanchetta, G., Jex, C., Wagner, B., Vogel, H. (2010): Late Quaternary palaeoenvironmental reconstruction from Lakes Ohrid and Prespa (Macedonia/Albania border) using stable isotopes, Biogeosciences, 7, 3109–3122.

Lézine, A. M., von Grafenstein, U., Andersen, N., Belmecheri, S., Bordon, A., Caron, B., Cazet, J. P., Erlenkeuser, H., Fouache, E., Grenier, C., Huntsman-Mapila, P., Hureau-Mazaudier, D., Manelli, D., Mazaud, A., Robert, C., Sulpizio, R., Tiercelin, J. J., Zanchetta, G., Z., Z. (2010): Lake Ohrid, Albania, provides an exceptional multi-proxy record of environmental changes during the last glacial-interglacial cycle, Palaeogeography, Palaeoclimatology, Palaeoecology, 287, 116–127.

Liermann, A. (2010a): Eine geologische Aufnahme der Umgebung von Kosel am Ohridsee (FYROM), Master's Thesis, University of Cologne.

Liermann, A. (2010b): Exhumierungsgeschichte der Region am Ohrid- und Prespasee in Mazedonien (FYROM) und Albanien durch Spaltspur-Thermochronologie an Apatit, Master's Thesis, University of Cologne.

Lindhorst, K., Vogel, H., Krastel, S., Wagner, B., Hilgers, A., Zander, A., Schwenk, T., Wessels, M., Daut, G. (2010): Stratigraphic analysis of Lake-level Fluctuations in Lake Ohrid: An integration of high resolution hydro-acoustic data and sediment cores, Biogeosciences, 7, 3531–3548.

Lindhorst, K., Gruen, M., Krastel, S., Schwenk, T. (2012a): Hydroacoustic Analysis of Mass Wasting Deposits in Lake Ohrid (FYR Macedonia/Albania), vol. Submarine Mass Movements and Their Consequences, 245 of *Advances in Natural and Technological Hazards Research*, Springer Science+Business Media.

Lindhorst, K., Krastel, S., Reicherter, K., Stipp, M., Wagner, B., Schwenk, T. (2012b): Sedimentary and tectonic evolution of Lake Ohrid (Macedonia/Albania), Basin Research.

Lowrie, W. (2007): Fundamentals of Geophysics, Cambridge University Press, 2nd edn.

Malte-Brun, M. (1929): Universal Geography.

Marrett, R., Peacock, D. . (1999): Strain and stress, Journal of Structural Geology, 21, 1057–1063.

Martens, K. (1997): Speciation in ancient lakes, Trends in Ecology and Evolution, 12 (5), 177–182, doi:10.1016/S0169-5347(97)01039-2.

Martin, P. (1994): Lake Baikal, in: Speciation in Ancient Lakes, edited by Martens, K., Goddeeris, B., Coulter, G., vol. 44, pp. 3–11, Schweizerbart.

Matzinger, A., Jordanoski, M., Veljanoska-Sarafiloska, E., Sturm, M., M. B., Wüest, A. (2006a): Is Lake Prespa jeopardizing the ecosystem of ancient Lake Ohrid?, Hydrobiologia, 553, 89–109.

Matzinger, A., Spirkovski, Z., Patceva, S. Wüest, A. (2006b): Sensitivity of ancient Lake Ohrid to local anthropogenic impacts and global warming, Journal of Great Lakes Research, 32, 158–179.

Matzinger, A., Schmid, M., Veljanoska-Sarafiloska, E., Patceva, S., Guseka, D., Wagner, B., M., S., Mller, B., Wüest, A. (2007): Assessment of early eutrophication in ancient lakes - A case study of Lake Ohrid, Limnology and Oceanography, 52, 338–353.

McCalpin, J. P. (2009): Paleoseismology, Academic Press Inc, San Diego, 2nd edn.

McClay, K., Ellis, P. (1987): Geometries of extensional fault systems developed in model experiments, Geology, 15, 341–344.

McClusky, S., Balassanian, S., Barka, A., Demir, C., Erginav, S., Georgiev, I., Gurkan, O.,

Hamburger, M., Hurst, K., Kahle, H., Kastens, K., Kekelidze, G., King, R., Kotzev, V., Lenk, O., Mahmoud, S., Mishin, A., Nadarya, M., Ouzounis, A., Paradissis, D., Peter, Y., Prilepin, M., Reilinger, R., Sanli, I., Seeger, H., Tealeb, A., Toksoz, M., Veis, G. (2000): Global Positioning System constraints on plate kinematics and dynamics in the eastern Mediterranean and Caucasus, Journal of Geophysical Research, 105, 5695–5719.

McKenzie, D. (1972): Active Tectonics of the Mediterranean Region, Geophysical Journal of the Royal Astronomical Society, 30, 109–185.

Mercier, J., Carey, E., Phillip, H., Sorel, D. (1976): La neotectonique plio-quaternaire de lÁrc Egeen externe et de la Mer Egée et ses relations avec séismicité, Le Bulletin de la Société géologique de France, 18, 355–372.

Michetti, A., Brunamonte, F., Serva, L., Whitney, R. (1995): Seismic hazard assessment from paleoseismological evidence in the Rieti region (Central Italy), Bullettin of the Asssociation of Engineering Geologists, Spec. Publ., 6 (Perspectives in Paleoseismology), pp. 63–82.

Michetti, A. M., Hancock, P. L. (1997): Paleoseismology: Understanding past earthquakes using Quaternary geology, Geodynamics, 24 1-4, 3–10.

Michetti, A. M., Audemard, F. A., Marco, S. (2005): Future trends in Paleoseismology: Integrated study of the seismic landscape as a vital tool in seismic hazard analyses, Tectonophysics, 408, 3–21.

Michetti, A. M., Esposito, E., Guerrieri, L., Porfido, S., Serva, L., Tatevossian, R., Vittori, E., Audemard, F., Azuma, T., Clague, J., Comerci, V., Gürpinar, A., McCalpin, J., Mohammadioun, B., Mörner, N.-A., Ota, Y., Roghozin, E. (2007): Intensity Scale ESI 2007, Memorie Descrittive della Carta Geologica D'Italia, 74, 1–41.

Mikulić, F. (1961): Neue Arten Candona aus dem Ohridsee, Bulletin du Museúm dú histoire naturelle (Beograd), Serie B, 17, 87–107.

Mikulić, F., Pljakić, M. (1970): Die Merkmale der kvalitativen Distribution der endemischen Candonaarten im Ochridsee, Ekologija, 5, 101–115.

Milivojevic, M. G. (1993): Geothermal model of earths crust and lithosphere for the territory of Yugoslavia: some tectonic implications, Studia Geophysica Et Geodaetica, 37, 265–278.

Milutinovic, Z. V., Trendafiloski, G., Olumceva, T. (1995): Disaster preparedness planning for small and medium size hospitals basedon structural, nonstructural and functional vulnerability assessment, World Health Organisation, Institute of Earthquake Engineering and Engineering Seismology, Skopje.

Muceku, B., Mascle, G., Tashko, A. (2006): First results of fission-track thermochronology in the Albanides, Tectonic Development of the Eastern Mediterranean Region, Geological Society of London Special Publications. edited by: Robertson, A. H. F. and Mountrakis, D., 260, 539–556.

Muceku, B., van der Beek, P., Bernet, M., Reiners, P., Mascle, G., Tashko, A. (2008): Thermochronological evidence for Mio- Pliocene late orogenic extension in the northeastern Albanides (Albania), Terra Nova, 20, 180–187.

Muco, B. (1998): Catalogue of $M_L \leq 3.0$ earthquakes in Albania from 1976 to 1995 and distribution of seismic energy released, Tectonophysics, 292, 311–319.

Mullins, C. E. (1977): Magnetic susceptibility of the soil and its significance in soil science - a review, Journal of Soil Science, 28, 223–246.

Murray, A., Wintle, A. (2000): Luminescence dating of quartz using an improved single-aliquot regenerative-dose protocol., Radiation Measurements, 32, 57–73.

Murray, A., Wintle, A. (2003): The single aliquot regenerative dose protocol: potential for improvements in reliability, Radiation Measurements, 37, 377–381.

NASA (2013): Landsat Data, Tech. rep., `http://landsat.gsfc.nasa.gov`, last access 2013-08-25.

Neal, A. (2004): Ground-penetrating radar and its use in sedimentology: principles, problems and progress, Earth Science Reviews, 66, 261–330.

NEIC (2013): USGS Earthquake Hazard Programme, National Earthquake Information Center, Tech. rep., `http://earthquake.usgs.gov/regional/neic/`, last access: 2013-10-03.

Palumbo, L., Benedetti, L., Bourles, D., Cinque, A., Finkel, R. (2004): Slip history of the Magnola Fault (Apennines, Central Italy) from 36Cl surface exposure dating: evidence for strong earthquakes over the Holocene, Earth and Planetary Science Letters, 225, 163–176.

Pamić, J., Gusić, I., V., J. (1998): Geodynamic evolution of the Central Dinarides, Tectonophysics, 297, 251–268.

Papanikolaou, D., Alexandri, M., Nomikou, P. (2006): Active faulting in the north Aegean basin, Postcollisional Tectonics and Magmatism in the Mediterranean Region and Asia, Geological Society of America, 409 (11), 189–209.

Papanikolaou, I., Papanikolaou, D. (2007): Seismic hazard scenarios from the longest geologically constrained active fault of the Aegean, Quaternary International, 171-172, 31–44.

Papanikolaou, I., Roberts, G., Michetti, A. (2005): Fault scarps and deformation rates in Lazio-Abruzzo, Central Italy: comparison between geological fault slip-rate and GPS data., Tectonophysics, 408, 147–176.

Papanikolaou, I. D., Foumelis, M., Parcharidis, I., Lekkas, E. F., Fountoulis, I. G. (2010): Deformation pattern of the 6 and 7 April 2009, MW= 6.3 and MW = 5.6 earthquakes in L'Aquila (Central Italy) revealed by ground and space based observations, Natural Hazards and Earth System Science, 10, 73–87.

Papazachos, B., Papazachou, K. (1997): The earthquakes of Greece, Thessaloniki Editions Ziti.

Peters, R. (2010): Felskartierung und strukturelle Aufnahme bei Ohrid (FYROM), Master's Thesis, RWTH Aachen University.

Petkovski, T. (1959): Beitrag zur Kenntnis der Ostracoden-Fauna Jugoslawiens (V)., Publications of the Hydrobiological Research Institute, Faculty of Science, University of Istanbul, Seri B 4(4), 158–165.

Petkovski, T. (1960a): Süsswasserostracoden aus Jugoslavien VII., Fragmenta Balcanica, Musei Macedonici Scientiarum Naturalium, 3(12), 99–108.

Petkovski, T. (1960b): Zur Kenntnis der Crustaceen des Prespasees., Fragmenta Balcanica, Musei Macedonici Scientiarum Naturalium, 3, 117–131.

Petkovski, T. (1960c): Zwei neue Ostracoden aus dem Ohrid- und Prespasee., Izdanija Institut de Pisciculture de la R. P. Macedonie, 3, 57–66.

Petkovski, T. (1969a): Einige neue und bemerkenswerte Candoninae aus dem Ohridsee und einigen anderen Fundorten in Europa., Musei Macedonici Scientiarum Naturalium, 11, 81–111.

Petkovski, T. (1969b): Zwei neue Limnocythere-Arten aus Mazedonien (Crustacea-Ostracoda)., Musei Macedonici Scientiarum Naturalium, 12, 1–18.

Petkovski, T., Keyser, D. (1992): Leptocythere ostrovskensis sp. n. (Crustacea, Ostracoda, Cytheracea) aus dem See Vegoritis (Ostrovsko Ezero) in NW Griechenland. Mit kurzer bersicht der Süsswasserarten des Genus Leptocythere G.O. Sars, 1925 vom Westbalkan., Mitteilungen Hamburgisches Zoologisches Museum und Institut, 89, 227–237.

Petrovski, J. T. (2004): Damaging effects of July 26, 1963 Skopje Earthquake, Tech. rep., Middle East Seismological Forum (MESF), `www.meseisforum.net`, last access: 15 June 2010.

Popovska, C., Bonacci, O. (2007): Basic data on the hydrology of Lakes Ohrid and Prespa, Hydrol. Proc., 21, 658–664.

Premti, I., Dobi, A. (1994): A geological map of Albania, Sheet Masivi Ultrabazik i Shebenik-Pogradecit, Inst. Stud. Proj. Gjeologjike, Tirana.

Ramsay, J., Lisle, R. (2000): The Techniques of Modern Structural Geology, Volume 3: Applications of continuum mechanics in structural geology, Academic Press, London, pp. 701–1061.

Reches, Z. (1987): Determination of the tectonic stress tensor from slip along faults that obey the Coulomb yield condition, Tectonics, 6(6), 849–861.

Reicherter, K., Peters, G. (2005): Neotectonic evolution of the Central Betic Cordilleras (Southern Spain), Tectonophysics, 405, 191–212.

Reicherter, K., Jabaloy, A., Galindo-Zaldívar, J., Ruano, P., Becker-Heidmann, P., Morales, J., Reiss, S., González-Lodeiro, F. (2003): Repeated palaeoseismic activity of the Ventas de Zafarraya fault (S Spain) and its relation with the 1884 Andalusian earthquake, Int J Earth Sci (Geol Rundsch), 92, 912–922.

Reicherter, K., Michetti, A. M., Silva, P. G. (2009): Introduction, in: Paleoseismology: Historical and prehistorical records of earthquake ground effects for seismic hazard assessment, edited by Reicherter, K., Michetti, A. M., Silva, P. G., vol. 316, pp. 1–10, The Geological Society, London, Special Publications.

Reicherter, K., Hoffmann, N., Lindhorst, K., Krastel-Gudegast, S., Fernández-Steeger, T., Wiatr, T. (2011): Active basins and neotectonics: morphotectonics of the Lake Ohrid Basin (FYROM/Albania), Z. dt. Ges. Geowiss., 162/2, 217–234.

Reiter, F., Acs, P. (2012): TectonicsFP - Kargl Computer & Consulting, `http://www.tectonicsfp.com/`.

Roberts, G. (1996): Variation in fault-slip directions along active and segmented normal fault systems, Journal of Structural Geology, 18, 835–845.

Roberts, G., Ganas, A. (2000): Fault-slip directions in central-southern Greece measured from striated and corrugated fault planes: comparison with focal mechanism and geodetic data, Journal of Geophysical Research, 105,23, 443–462.

Robertson, A. (2004): Development of concepts concerning the genesis and emplacement of Tethyan ophiolites in the Eastern Mediterranean and Oman regions, Earth Sci. Rev., 66, 331–387.

Robertson, A., Shallo, M. (2000): Mesozoic-Tertiary tectonic evolution of Albania in its regional Eastern Mediterranean context, Tectonophysics, 316, 197–254.

Rothwell, R., Rack, F. (2006): New Techniques in sediment core analysis: an introduction, Geological Society London, Special Publications, 267, 1–29.

Sadler, M. (2010): Morphologie einiger Abschiebungen des Ohrid Beckens (EJR Mazedonien/Albanien), bachelor's Thesis, RWTH Aachen University (2010).

Sandmeier, K. H. (2010): Reflex-Win Version 5.6 radar processing and interpretation software package, Sandmeier Scientific Software, Karlsruhe, Germany, `http://www.sandmeier-geo.de`.

Schmid, S., Bernoulli, D., Fügenschuh, B., Matenco, L., Schefer, S., Schuster, R., Tischler, M., Ustaszewski, K. (2008): The Alpine-Carpathian-Dinaridic orogenic system: correlation and evolution of tectonic units, Swiss Journal of Geoscience, 101, 139–183, doi:10.1007/s00015-008-1247-3.

Schmid, S., Bernoulli, D., Fügenschuh, B., Kounov, A., Matenco, L., Oberhänsli, S., Schefer, S., Ustaszewski, K., van Hinsbergen, D. (2012): Tectonic units of the Alpine Collision Zone between Eastern Alps and Western Turkey, Tech. rep., Institute of Geology and Paleontology University Basel, `http://pages.unibas.ch/earth/tecto/`, last access: 2012-06-04.

Serva, L. (1995): Criteri geologici per la valutazione della sismicit: considerazioni e proposte, Atti dei Convegni Lincei (Terremoti in Italia), 122, 103–116, academia Nazionale dei Lincei Roma.

Shimaraev, M., Verbolov, V., Granin, N., Sherstayankin, P. (1994): Physical limnology of Lake Baikal: a review, Baikal International Center fo Ecological Research.

SIAL (2013): Species in Ancient Lakes, Tech. rep., http://www.sial-online.org, last access 2012-04-09.

Sippel, J. (2008): The Paleostress History of the Central European Basin System, Ph.D. Thesis, University of Berlin.

Sippel, J., Scheck-Wenderoth, M., Reicherter, K., Mazur, S. (2009): Paleostress states at the south-western margin of the Central European Basin System - application of fault-slip analysis to unravel a polyphase deformation pattern, Tectonophysics, 470, 129–146.

Sippel, J., Saintot, A., Heeremans, M., Scheck-Wenderoth, M. (2010): Paleostress field reconstruction in the Oslo region, Marine and Petroleum Geology, 27, 682–708.

Spang, J. (1972): Numerical Method for dynamic Analysis of Calcite Twin Lamellea, Geological Society America, Bulletin, 83, 467–472.

Sperner, B. (1996): Computer programs for the kinematic analysis of brittle deformation structures and the Tertiary tectonic evolution of the Western Carpathians (Slovakia), in: Tbinger Geowissenschaftliche Arbeiten (TGA), Institut und Museum fr Geologie und Palontologie der Universitt Tbingen, vol. 27, p. 120.

Sperner, B., Ratschbacher, L., Ott, R. (1993): Fault-striae analysis: a turbo pascal program package for graphical presentation and reduced stress tensor calculation, Computers and Geosciences, 19, 1361–1388.

Stankovic, S. (1960): The Balkan Lake Ohrid and Its Living World, Monographiae Biologicae IX, Uitgeverij Dr. W. Junk b.v. Publishers, The Hague, p. 357 pp.

Stewart, I. (1993): Sensitivity of fault-generated scarps as indicators of active tectonism: some constraints from the Aegean region, in: Landscape Sensitivity, edited by Thomas, D., Allison, R., pp. 129–147, Wiley, Chicheste.

Stewart, I., Hancock, P. (1994): Neotectonics, Hancock, P.L. (ed.), Continental Deformation.

Stewart, I. S., Hancock, P. L. (1990): What is a fault scarp?, Episodes, 13, 256–263.

Stojardinovic, C. (1969): Multi-year oscillation of the lake-levels of Ohrid and Prespa lakes (in Macedonian language), Department of Mathematics and Natural Science, Skopje. Unpublished Maps (1969).

Stuiver, M., Polach, H. (1977): Discussion: Reporting of 14C Data, Radiocarbon, 19, 355–356.

Tari, V. (2002): Evolution of the northern and western Dinarides: a tectonostratigraphic approach, EGU Stephan Mueller Special Publication Series European Geosciences Union, 1, 223–236.

Thatcher, W., Foulger, G., B.R., J., Svarc, J., Quilty, E., G.W., B. (1999): Present-Day Deformation Across the Basin and Range Province, Western United States, Science, 283 (5408), 1714–1718, doi:10.1126/science.283.5408.1714.

Trajanovski, S., Wilke, T., Budzakoska-Djoreska, B., Krstanovski, Z. (2006): Evolution of the Dina Ancient Lake Species Flock (Hirudinae: Erpobdellidae) in lake Ohrid, Berliner Paläobiologische Abhandlungen, 9(63).

Tremblay, A., Meshi, A., Bédard, J. H. (2009): Oceanic core complexes and ancient oceanic lithosphere: insights from Iapetan and Tethyan ophiolites (Canada and Albania), Tectonophysics, 473, 36–52.

Twiss, R., Moores, E. M. (2007): Structural Geology, W.H. Freemann and Company; New York.

UNESCO (2013): UNESCO World Heritage Center, Tech. rep., `http://whc.unesco.org/`, last access 2013-10-01.

USGS (2012): Aster data access, Tech. rep., `http://lpdaac.usgs.gov/`, last access 2013-09-23.

Vittori, E., Serva, L., Sylos Labini, S. (1991): Palaeoseismology: review of the state-of-the-art., Tectonophysics, 193, 9–32.

Vogel, H., Wagner, B., Zanchetta, G., Sulpizio, R., Rosén, P. (2010a): A paleoclimate record with tephrochronological age control for the last glacial-interglacial cycle from Lake Ohrid, Albania and Macedonia, Journal of Paleolimnology, 44, 295–310, doi:10.1007/s10933-009-9404-x.

Vogel, H., Zanchetta, G., Sulpizio, R., Wagner, B., Nowaczyk, N. (2010b): A tephrostratigraphic record for the last glacial-interglacial cycle from Lake Ohrid, Albania and Macedonia, Journal of Quaternary Science, 24, 1–19.

Wagner, B., Wilke, T. (2011): Evolutionary and geological history of the Balkan lakes Ohrid and Prespa, Biogeosciences, 8, 995–998.

Wagner, B., Reicherter, K., Daut, G., Wessels, M., Matzinger, A., Schwalb, A., Spirkovski, Z., Sanxhaku, M. (2008): The potential of Lake Ohrid for long-term palaeoenvironmental reconstructions, Palaeogeography, Palaeoclimatology, Palaeoecology, 259, 241–356.

Wagner, B., Vogel, H., Zanchetta, G., Sulpizio, R. (2010): Environmental change within the Balkan region during the past ca. 50 ka recorded in the sediments from lakes Prespa and Ohrid, Biogeosciences, 7, 3187–3198.

Wagner, B., Francke, A., R., S., Zanchetta, G., Lindhorst, K., Krastel, S., Vogel, H., Daut, G., Grazhdani, A., B., L., Trajanovski, S. (2012): Seismic and sedimentological evidence 1 of an early 6th century AD earthquake at Lake Ohrid (Macedonia/Albania).

Wallace, R. (1951): Geometry of shearing stress and relation to faulting, Journal of Geology, 59(2), 118–130.

Walter, M. (2009): Felskartierung entlang der Küste des Ohridsees (Mazedonien) zwischen Gradiste (Autocamp) und Sveti Naum, Master's Thesis, RWTH Aachen University.

Watzin, M. C. Puka, V., Naumoski, T. B. (2002): Lake Ohrid and its Watershed, State of the Environment Report, Tech. rep., Lake Ohrid Conservation Project. Tirana, Albania and Ohrid, Macedonia.

Wells, D. L., Coppersmith, J. K. (1994): New empirical relationships among magnitude, rupture length, rupture width, rupture area, and surface displacement, Bulletin of the Seismological Society of America, 84, 974–1002.

Wernicke, B. (1981): Low-angle normal faults in the Basin and Range Province: nappe tectonics in an extending orogen, Nature, 291(5817), 645–648.

Wernicke, B. & Burchfiel, B. C. (1982): Modes of extensional tectonics, Journal of Structural Geology, 4(2), 105–115.

Wilke, T., Albrecht, C., Anistratenko, V., Sahin, S., Yildirim, M. (2007): Testing biogeographical hypotheses in space and time: Faunal relationships of the putative ancient lake Egirdir in Asia Minor., Journal of Biogeography, 34, 1807–1821.

Wintle, A., Murray, A. (2006): A review of quartz optically stimulated luminescence characteristics and their relevance in single-aliquot regeneration dating protocols, Radiation Measurements, 41, 369–391.

Wu, J., McClay, K., Whitehouse, P., Dooley, T. (2009): 4d Analogue Modelling of Transtensional Pull-Apart Basins, Marine and Petroleum Geology, 26, 1608–1623.

Yamaji, A. (2000): The multiple inverse method: a new technique to separate stresses from heterogeneous fault-slip data, Journal of Structural Geology, 22, 441–452.

Yamaji, A. (2007): An Introduction to Tectonophysics-Theoretical Aspects of Structural Geology, TERRAPUB.

Yanev, Y., Boev, B., Doglioni, C., Innocenti, F., Manetti, P., Pecskay, Z., Tonarini, S., D´ Orazio, M. (2008): Late Miocene to Pleistocene potassic volcanism in the Republic of Macedonia, Mineralogy and Petrology, 94, 45–60, doi:10.1007/s00710-008-0009-2.

A. Appendix

Figure A.1. ***(facing page)*:** PBT and tangent lineation plots of east coast outcrop data. The PBT-axes plots of the complete dataset per location is plotted in the first column; Columns 2-5 distribute the tangent lineation plots of each subset with mean σ_1 (triangle) and σ_3 (star) angles and color coded misfit angles distribution; the last column shows the rest that could not be assigned to any of the stress states.

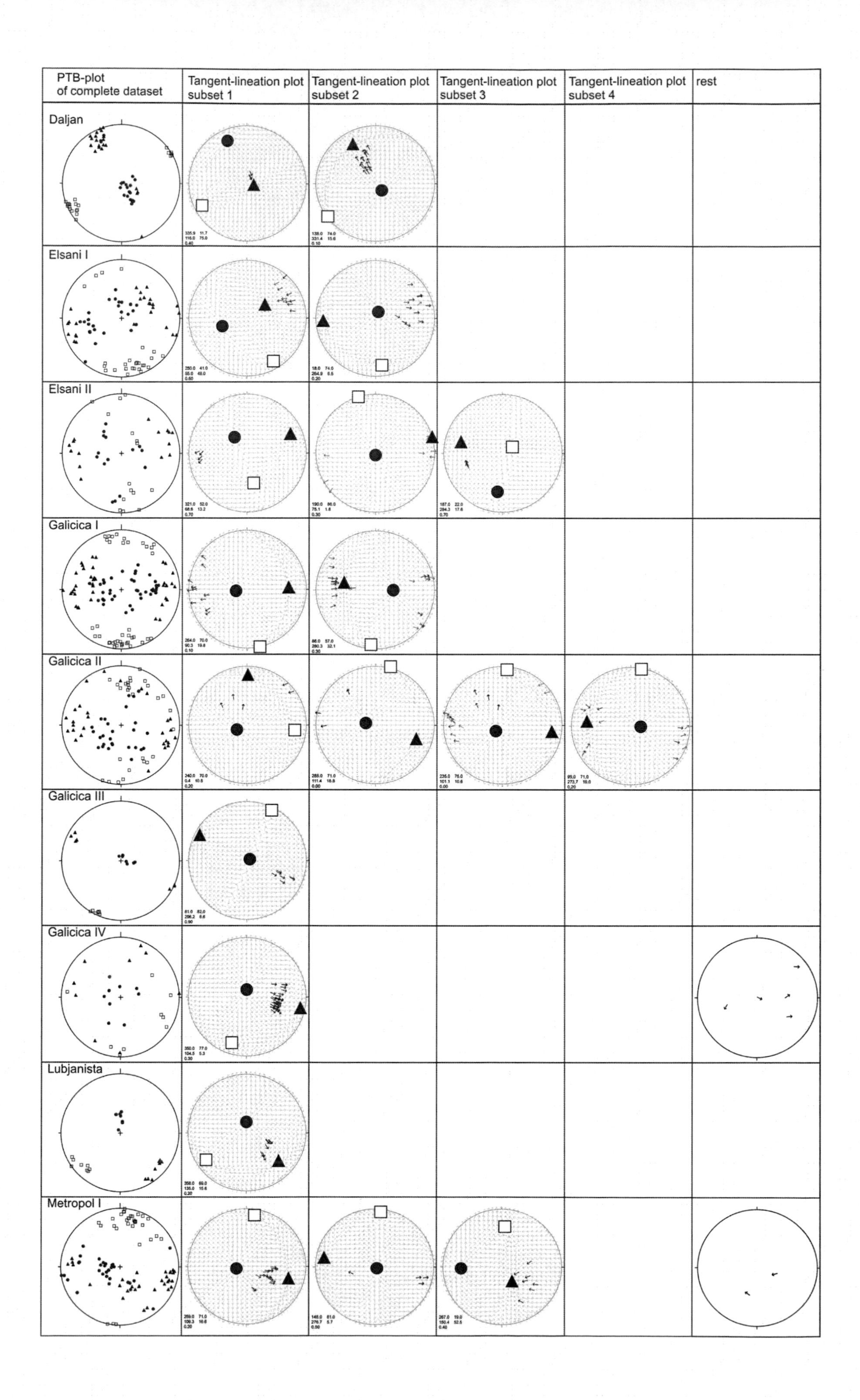

PTB-plot of complete dataset
Tangent-lineation plot subset 1
Tangent-lineation plot subset 2
Tangent-lineation plot subset 3
Tangent-lineation plot subset 4
rest
Daljan
Elsani I
Elsani II
Galicica I
Galicica II
Galicica III
Galicica IV
Lubjanista
Metropol I

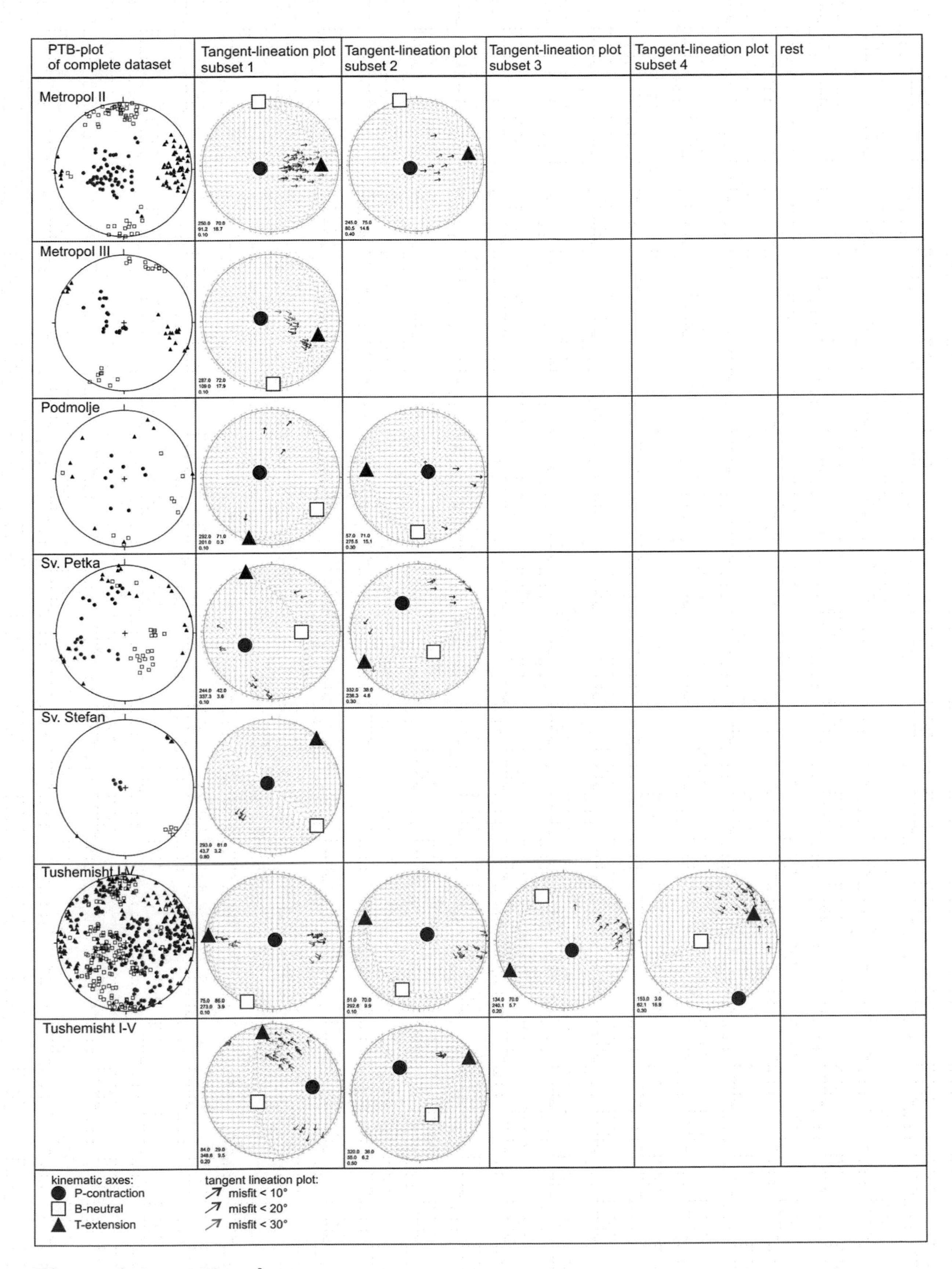

Figure A.1.: continued

Figure A.2. ***(facing page):*** PBT and tangent lineation plots of west coast outcrop data. The PBT-axes plots of the complete dataset per location is plotted in the first column; Columns 2-5 distribute the tangent lineation plots of each subset with mean σ_1 (triangle) and σ_3 (star) angles and color coded misfit angles distribution; the last column shows the rest that could not be assigned to any of the stress states.

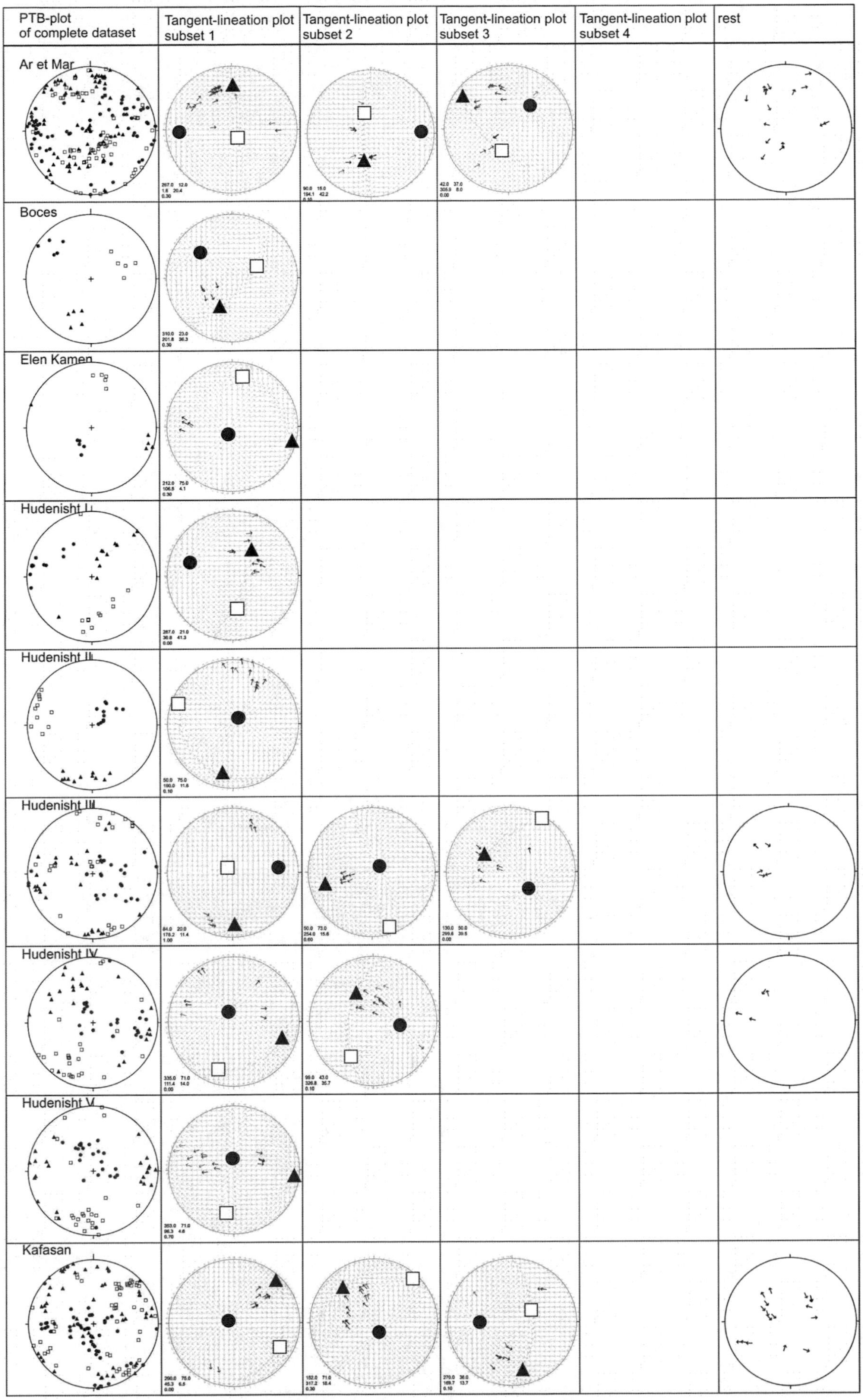

PTB-plot of complete dataset	Tangent-lineation plot subset 1	Tangent-lineation plot subset 2	Tangent-lineation plot subset 3	Tangent-lineation plot subset 4	rest
Ar et Mar	267.0 12.0 1.6 20.4 0.30	90.0 15.0 194.1 42.2 0.10	42.0 37.0 305.9 8.0 0.00		
Boces	310.0 23.0 201.8 36.3 0.30				
Elen Kamen	212.0 75.0 106.5 4.1 0.30				
Hudenisht I	287.0 21.0 36.8 41.3 0.00				
Hudenisht II	50.0 75.0 190.0 11.6 0.10				
Hudenisht III	84.0 20.0 178.2 11.4 1.00	50.0 73.0 254.0 15.6 0.60	130.0 50.0 299.6 39.5 0.00		
Hudenisht IV	335.0 71.0 111.4 14.0 0.00	99.0 43.0 326.8 35.7 0.10			
Hudenisht V	353.0 71.0 96.3 4.6 0.70				
Kafasan	290.0 75.0 45.3 6.5 0.00	152.0 71.0 317.2 18.4 0.30	270.0 36.0 169.7 13.7 0.10		

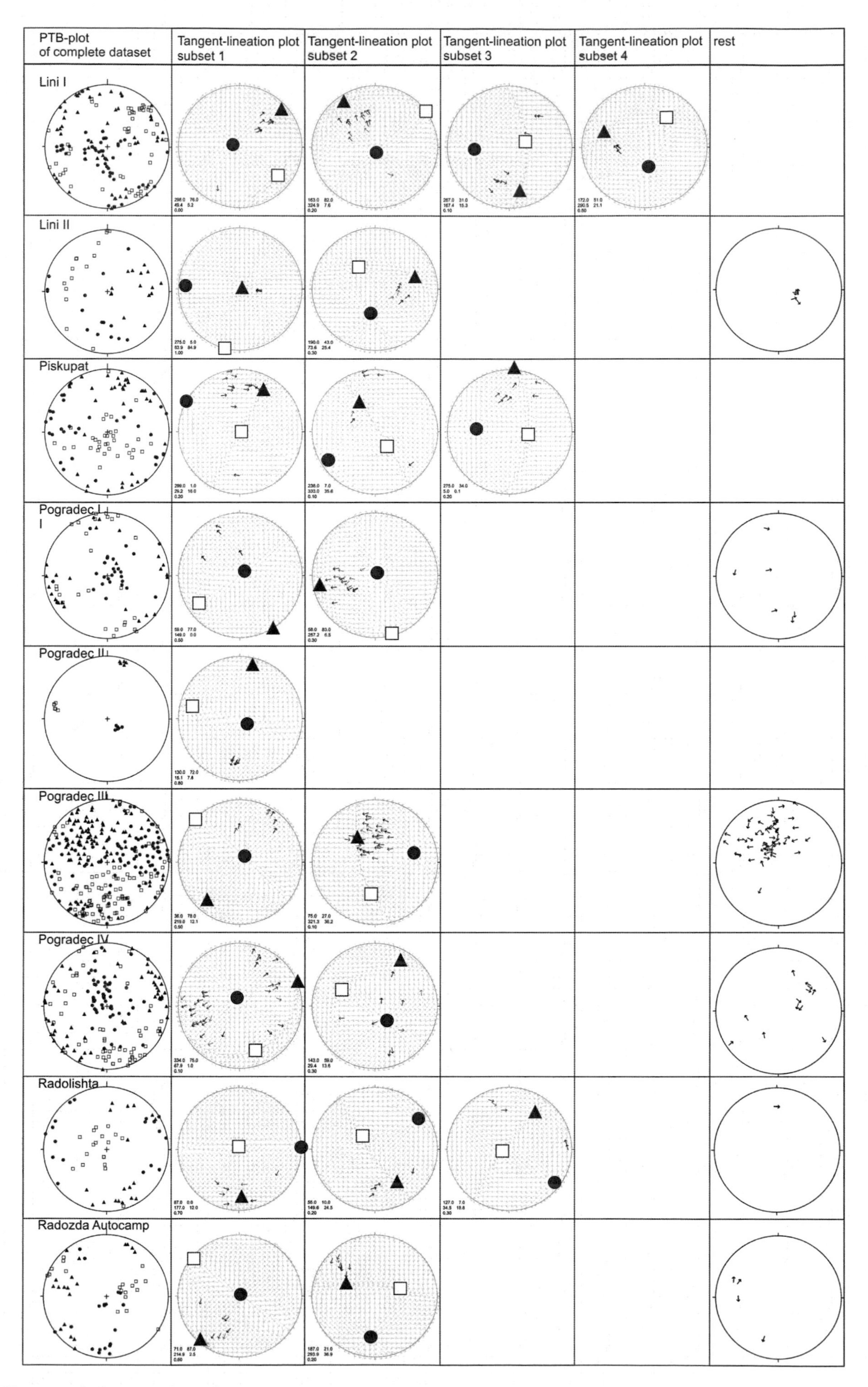

Figure A.2.: continued

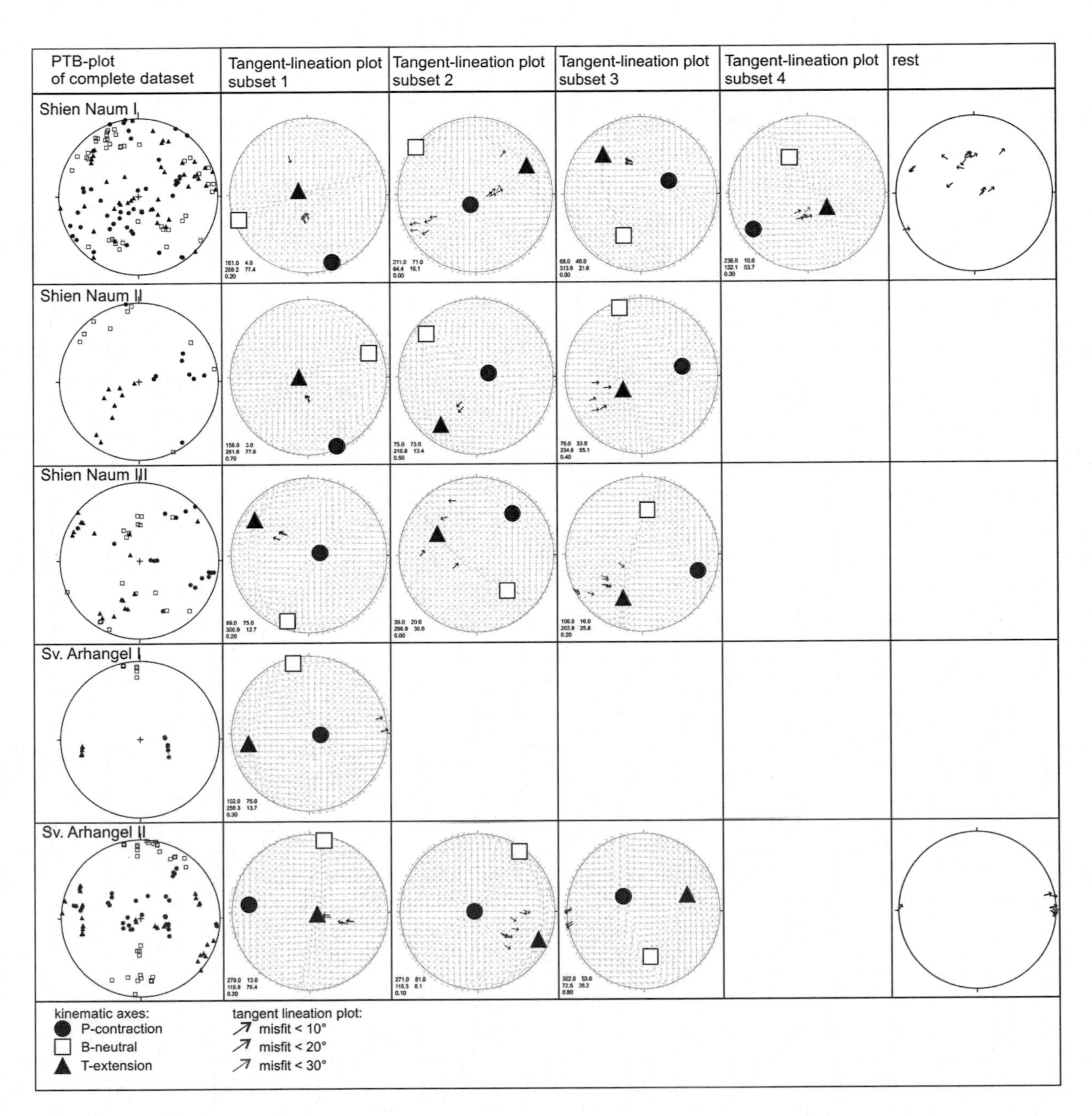

Figure A.2.: continued